Lithium Niobate Heterostructures

Heterostructures

Synthesis, properties and electron phenomena

Lithium Niobate-Based Heterostructures

Synthesis, properties and electron phenomena

Maxim Sumets

Department of Physics and Astronomy, The University of Texas Rio Grande Valley, USA

IOP Publishing, Bristol, UK

ISBN   978-0-7503-1729-0 (ebook)
ISBN   978-0-7503-1727-6 (print)
ISBN   978-0-7503-1728-3 (mobi)

DOI   10.1088/978-0-7503-1729-0

Version: 20180801

IOP Expanding Physics
ISSN 2053-2563 (online)
ISSN 2054-7315 (print)

British Library Cataloguing-in-Publication Data: A catalogue record for this book is available from the British Library.

Published by IOP Publishing, wholly owned by The Institute of Physics, London

IOP Publishing, Temple Circus, Temple Way, Bristol, BS1 6HG, UK

US Office: IOP Publishing, Inc., 190 North Independence Mall West, Suite 601, Philadelphia, PA 19106, USA

Contents

Preface

Ferroelectric materials are widely used in memory and high speed integrated optics devices. These materials have attracted growing interest due to their unique optoacoustic and optoelectric properties including nonvolatility, fast switching and radiative stability. Ferroelectric thin films are especially attractive due to the possibility of integrating them into semiconductor and optoelectronic devices. Remarkably good optical localization in the waveguide geometry allows the non-linear optical properties of these materials to be utilized.

Lithium niobate ($LiNbO_3$) being a ferroelectric material, with high Curie temperature (T_c = 1210 °C), wide band gap and high piezoelectric and electro-optic coefficients, is one of the most promising ferroelectric materials with interesting nonlinear optical properties.

The possibility of integration of lithium niobate into various substrates led to the development and fabrication of versatile integrated optical devices such as non-volatile memory devices, Mach–Zehnder modulators, surface-acoustic wave (SAW) filter modulators, frequency filters, microring resonators, optical amplitude modulators etc.

In recent years there has been a tendency to substitute electronic devices with photonic ones in terms of optical information transition. Compared to electronic devices, photonic systems offer a range of advantages such as an operation based on wide band gap materials, the absence of electromagnetic interference, possible photon multiplication etc. Therefore, several attempts have been made to fabricate integrated photonics devices using the silicon on insulator (SOI) approach. In contrast to SOI, a technology based on lithium niobate offers perfect electro-optic, acousto-optic and nonlinear optic properties. Therefore, $LiNbO_3$-based photonic elements (photonic wires for instance) can enforce development of integrated devices with active elements such as optical modulators, tunnel filters, nonlinear wavelength converters and various tunnel lasers. Fabrication of thin $LiNbO_3$ films with a high degree of crystallinity is a key point in this issue. The use of substrates with the required properties restricts the synthesis possibility of thin epitaxial films due to significant mismatch of the crystal lattices. Consequently, the synthesis of thin $LiNbO_3$ films with a high quality $LiNbO_3$/substrate heterointerface is a highly actual issue.

The radio-frequency magnetron sputtering (RFMS) method and the ion beam sputtering (IBS) method are two of the most effective deposition methods in vacuum technologies for thin films of complex oxides preserving the initial elemental composition and offering better adhesion to a substrate. However, the high sensitivity of film properties to technological parameters of RFMS leads to broad possibilities in the fabrication of films with the required parameters, but on the other hand leads to some problems with reproducibility of these properties. The only way to solve this problem is to develop and keep up with technological regimes of RFMS ensuring the growth of $LiNbO_3$ films with the required structural, electrical, ferroelectric and optical properties. The complexity of this issue is the wide range of critical parameters, influencing the stable fabrication of LiNbO$_3$ with the properties listed above. Despite the numerous papers published in the past two decades on the synthesis of LiNbO$_3$ films, most of them dealt with the impact of technological parameters on the structure, composition and morphology of fabricated films. At the same time, the electrical properties of LiNbO$_3$ film depending on fabrication regimes are not so widely reflected in the literature because this influence is not so obvious and clear. Unfortunately, electrical properties of LiNbO$_3$ reported in literature are complementary to structural studies due to the absence of systematic analysis of electronic phenomena in LiNbO$_3$-based heterostructures as an elemental basis of contemporary electronics.

Taking into account all that has been mentioned above, the development of RFMS technological fabrication regimes of LiNbO$_3$-based heterostructures with the required structural and electrical properties is crucial for practical applications and from a fundamental point of view.

This work is aimed at characterizing the substructure and electronic properties of a heterosystem, formed in the deposition process of LiNbO$_3$ films onto the surfaces of silicon wafers.

The focus of this book

For applied purposes it is motivated by the point of view of the fabrication of metal–insulator–semiconductor (MIS) heterostructures—the functional elements of memory units and integrated optoelectronics devices.

From a fundamental viewpoint it is motivated by the lack of systematic data regarding substructure and electronic properties of a heterosystem, formed in the growth process of LiNbO$_3$ films on the surface of silicon wafers during the radio-frequency magnetron sputtering process.

This monograph is based on many years of study conducted by the author in the field of LiNbO$_3$-based heterostructures, thin films, electronic materials and materials science.

The scientific novelty of this work is reflected in the following study results:

1. The patterns of structural changes during the growing process of thin films deposited by RFMS and IBS methods have been revealed.
2. The structural transitions in the process of thermal annealing of as-grown films fabricated by RFMS method have been revealed.
3. The effect of deposition conditions and the spatial inhomogeneity of plasma on structure, composition and surface morphology of LiNbO$_3$ films have been systematically studied.
4. Based on the conducted electrical measurements and the proposed energy band diagram *at the first time* the charge transport mechanism in Si–LiNbO$_3$ heterostructures was studied in detail. It was revealed that at weak electric fields ($0 < E < 2$ kV cm^{-1}) and at room temperature hopping conductivity over charge centers in the band gap of LiNbO$_3$ takes place. In the range of moderate fields ($2 < E < 30$ kV cm^{-1}) Schottky emission and Fowler–Nordheim tunneling caused by the presence of charge centers 1.7 eV below the conduction band in LiNbO$_3$ are the prevalent mechanisms. The charge center concentration is affected by the deposition method and has a maxima

for films deposited by IBS ($N_t \sim 3 \cdot 10^{19}$ cm^{-3}). At strong fields ($E > 30$ kV cm^{-1}) conductivity obeys thermally-assisted tunneling through grain boundaries with potential barrier of $\varphi_b = 0.7$ eV.

5. Based on the study of electrical properties, it has been proved that the positive oxide charge created in LiNbO$_3$ thin films fabricated by RFMS method in an Ar atmosphere, caused by intrinsic defects (antisite defects 'Nb at the place of Li': $Nb_{Li}{}^{4\bullet}$, forming the charge centers in LiNbO$_3$ 0.7 eV below the conduction band), and it is not caused by the defects created by bombardment of the film surface or the presence of a LiNb$_3$O$_8$ phase.

6. *For the first time* it has been established that lithium ions, diffusing in the Si substrate, form shallow donor levels with concentration of $N_d = 5 \times 10^{17}$ cm^{-3} at the Si/LiNbO$_3$ heterointerface and declining exponentially in the bulk.

7. Based on the study of charge center parameters, it has been confirmed that the use of an Ar + O$_2$ gas mixture as a reactive gas environment leads to a significant decrease in the positive oxide charge in LiNbO$_3$ films, improving electrical properties of Si–LiNbO$_3$ heterostructures.

8. The effectiveness of thermal annealing as a method of improvement of electrical properties of Si–LiNbO$_3$ heterostructures has been demonstrated, leading to a decrease in the positive oxide charge in LiNbO$_3$ films and the density of states at the Si/LiNbO$_3$ interface.

9. Si–LiNbO$_3$ heterostructures, changing their conductivity type from p- to n-type in the temperature range of 100–200 K have been fabricated by the RFMS method.

The book contains four chapters. Chapter 1 presents a contemporary overview of the main properties of bulk lithium niobate, the advantage of thin LiNbO$_3$ films and their applications in integrated electronics and optoelectronics. Various deposition techniques used for the fabrication of LiNbO$_3$ films are compared in terms of their advantages and disadvantages. The influence of RF magnetron sputtering parameters on structure and composition of LiNbO$_3$ films reported in the literature is discussed.

Chapter 2 presents original results regarding the influence of synthesis conditions of RFMS and the IBS methods on structure, composition and surface morphology of as-grown LiNbO$_3$ films. Optimal RFMS parameters are proposed for fabrication of the single phase c-oriented LiNbO$_3$ films and Si–LiNbO$_3$ heterostructures. Also, the possibility of fabrication of epitaxial LiNbO$_3$ films is demonstrated.

In chapter 3 the electrical properties of Si–LiNbO$_3$ heterostructures, fabricated by RFMS and IBS methods without the ion assist effect, are studied. Based on the current–voltage, capacitance–voltage characteristics an energy band diagram of fabricated Si–LiNbO$_3$ heterostructures is proposed. Charge transport mechanisms in the studied heterostructures are investigated in detail. Using impedance spectroscopy methodology, ac hopping conductivity mechanisms are studied. It is demonstrated that the dielectric relaxation is affected by the Maxwell–Wagner relaxation phenomenon.

Chapter 4 focusses on the effect of sputtering conditions and thermal annealing on electron phenomena in Si–LiNbO$_3$ heterostructures. It is shown that the spatial inhomogeneity of plasma, its composition and relative target–substrate position greatly influence electrical properties of Si–LiNbO$_3$ heterostructures. Charge centers existing in the band gap of LiNbO$_3$ films as well as electronic states at the Si–LiNbO$_3$ interface are extremely sensitive to the synthesis conditions. Thermal annealing (TA) of the studied films leads to an increased concentration of vacancies and interface states in Si–LiNbO$_3$ heterostructures with the effect of improving its electrical properties.

Scientific results, presented in this monograph have been reported in various conferences and symposiums and this is reflected in cited publications.

Acknowledgments

Without the outstandingly talented, collaborative and supportive research colleagues, the work covered in this book would not have been possible. Formulation of objectives, research planning and discussion were carried out jointly with Professor Valentin Ievlev (Voronezh State University, Russian Academy of Science). I would like to acknowledge Dr Alexander Kostuchenko (Voronezh State Technical University, Russia), Dr Gennadiy Kotov (Voronezh State University of Engineering Technologies, Russia), Dr Oleg Ovchinnikov (Voronezh State University, Russia), Professor Vladimir Shur (Ural Federal University, Russia), Dr Karen Martirosyan (University of Texas Rio Grande Valley, USA) and Mr Vladislav Dybov (Voronezh State Technical University, Russia) for valuable help in synthesis and measurements, associated with $LiNbO_3$-based heterostructures.

I would like to thank my brother Dr Pavel Sumets (University of Auckland, New Zealand) for helping me with editing this monograph.

My special thanks goes to Natalia Sumets, my wife, who would have liked me to spend more time with her, while I spent most of my free time in conducting research.

Author biography

Maxim Sumets

Maxim Sumets is a Lecturer in the Department of Physics and Astronomy at The University of Texas Rio Grande Valley, USA. His field of research is Materials Science with a focus on thin films, semiconductor heterostructures, ferroelectrics and their application. He obtained his Master and PhD degrees from the Voronezh State University, Russia, and has been actively involved in research and education for more than 20 years. His fields of research cover electrical and structural properties of materials.

Chapter 1

Thin films of lithium niobate: potential applications, synthesis methods, structure and properties

1.1 The structure and main properties of bulk lithium niobate

Lithium niobate crystals were grown for the first time by the Chokhralsky method in 1965 [1], and their structure was investigated by Abrahams in the series of works [2, 3]. Lithium niobate ($LiNbO_3$) belongs to a R3c space group, where the oxygens are arranged in nearly hexagonal close-packed planar sheets, as shown in figure 1.1.

The unit cell can be either rhombohedral (trigonal, $a = 5.4944$ Å, $\alpha = 55\,°52'$) or hexagonal ($a_H = 5.1483$ Å, $c_H = 13.8631$ Å, $c/a = 2.693$) with six formula units per unit cell. The stacking sequence of cations in these octahedral sites is: Nb, Li, vacancy, Nb, Li, vacancy, and so on [3]. The octahedral interstices formed in this structure are one-third filled by lithium atoms, one-third filled by niobium atoms, and one-third is vacant.

From this point of view bulk crystals with stoichiometric composition ($R = $ Li/Nb $= 1$) have a nearly ideal structure. In lithium reduced crystals, along with congruent crystals ($R = 0.946$), a cation sublattice is considerably disordered. It was demonstrated [4], that excessive niobium ions substitute lithium ions at their positions and, at the same time, a quite loose cation sublattice allows various ions to be introduced into the structure.

The distance of Li^+ to the nearest oxygen plane is 0.37 Å. The lattice asymmetry makes lithium niobate the polar material and as shown in figure 1.1, oxygen atoms reside in the oxygen triangles. To reverse the orientation of the polarization the Li^+ ion has to be displaced to the other side of the nearest oxygen triangle, whereas the Nb^{5+} ion can only move within the oxygen cages. Two stable positions (before and after reversal) define two possible directions of the spontaneous polarization.

For stoichiometric crystals the defects affected by the ordering of the cation sublattice play the most important role in the formation of optical characteristics.

doi:10.1088/978-0-7503-1729-0ch1

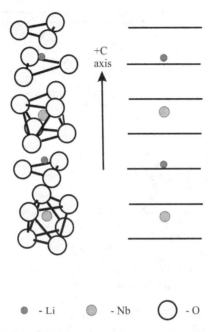

Figure 1.1. Positions of the lithium atoms and niobium atoms with respect to oxygen octahedra in the ferroelectric phase of lithium niobate. The horizontal lines on the right side of diagram represent the oxygen layers.

They are influenced by insignificant changes in the Li/Nb ratio or by the presence of a small amount of impurities capable of replacing Li and Nb ions and occupying the vacant octahedral voids causing local changes in the cation order along the polar axis. Impurity concentration of the order of hundredths of a percent can influence greatly the dielectric and optical properties of a crystal such as laser damage.

In the cation sublattice of lithium niobate the density inhomogeneity of the cluster type is observed changing the translational invariance of a structure without a change in the unit cell symmetry. Uniform composition along the crystal growth axis is not evidence that the composition of a crystal corresponds to the congruent melting conditions. The degree of crystal inhomogeneity can be increased greatly through growing them in an electric field or by thermal annealing near the melting point in a weak electric field. Lithium niobate single crystals, grown under congruent melting conditions have a disordered cation sublattice. They are very sensitive to laser damage, limiting their application in optical devices. There is no principal difference between congruent and stoichiometric crystals from the physical and chemical points of view—they differ only by degree of defectiveness.

At the Curie temperature of 1210 °C [5] $LiNbO_3$ has undergone the phase transition from the ferroelectric to the paraelectric state corresponding to the $R\bar{3}$ space group. Density of $LiNbO_3$ single crystals depends on the Li content and is equal to 4.648 g cm^{-3} for congruent $LiNbO_3$ [4].

The schematic phase diagram of the Li_2O–Nb_2O_5 system is shown in figure 1.2 [6].

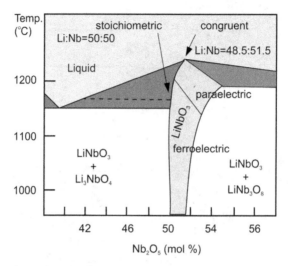

Figure 1.2. Schematic phase diagram of the Li_2O–Nb_2O_5 pseudobinary system near the congruent and stoichiometric composition of $LiNbO_3$ [6]. Reprinted by permission from Springer Nature, copyright 2007.

As can seen from figure 1.2, $LiNbO_3$ is a material with variable composition demonstrating a wide solid solution range. Specifically, there are areas, consisting of both $LiNbO_3$ oxide and either Li_3NbO_4, or $LiNb_3O_8$ which are centrosymmetric and hence non-ferroelectric phases. Therefore, $LiNbO_3$ single crystals are grown with special care at low temperatures, because a small deviation results in the formation of Li_3NbO_4, or $LiNb_3O_8$ oxides provoking degradation of ferroelectric properties.

With the aim, for example, of better grain boundary observation or formation of the required topology of thin film devices, chemical etching is sometimes of interest. Despite that lithium niobate is quite inert at room temperature, it was reported that $LiNbO_3$ single crystals are etched through boiling (110 °C) in the HF:HNO_3 (1:2) mixture [7]. Nevertheless, the etching of thin $LiNbO_3$ films by this method leads to their exfoliation from the substrate after a few minutes. In modern thin film technologies the most popular method is the reactive ion etching (RIE) ('dry') with the use of various reactive gas recipes. As regards lithium niobate, there is a wide range of etching approaches making use of various fluorine based and chlorine based plasmas such as CF4, CCl_2F_2, CHF_3 [8]. For instance, using the CF_4/Ar/H_2 reactive gas environment, it was revealed [9] that the etch rate and degree of surface crystallinity depend on time etching and flow rate of H_2. Other authors reported that depending on the compound ratio in their Ar/C_3H_8 reactive gas mixture, various degrees of etching anisotropy are attainable [10].

It is generally accepted, that the coefficient of thermal expansion is a critical parameter for understanding the lattice distortion. Due to symmetry, there are two linear coefficients of thermal expansion (longitudinal and transverse) with magnitudes for lithium niobate of $\alpha_{11} = 15.0 \times 10^{-6}$ and $\alpha_{33} = \alpha_{11} = 7.5 \times 10^{-6}$.

The design of memory units requires the dielectric permittivity to be studied and controlled. According to the axial symmetry the tensor of dielectric permittivity of lithium niobate has only two independent elements. Whereas the low frequency dielectric permittivity of bulk $LiNbO_3$ is $\varepsilon_{11} = 30$, $\varepsilon_{33} = 80$ [7].

The most attractive features of lithium niobate are its optical properties and electro-optical effect. The bulk lithium niobate is an optically transparent material in the range of wavelengths from 0.35 μm to 5 μm where the lattice absorption is observed. In spite of the fact that optical properties of lithium niobate are well established, the optical band gap, reported in the literature, varies significantly. For example, theoretical and experimental works showed that the optical band gap varies in the range between 3.57 eV [11] and 4.7 eV [12]. Some authors argue that this phenomenon can be caused by the presence of defects in $LiNbO_3$, generating localized states (even the band tails) influencing the absorption in single crystals. Lithium niobate manifests birefringence with the principal ordinary and extraordinary refractive indexes $n_o = 2.289$ and $n_e = 2.201$, respectively ($\lambda = 633$ nm). The refractive indexes depend on the Li/Nb ratio and vary significantly depending on composition [13].

Data, regarding the electrical properties of lithium niobate reported in the literature are controversial and affected by the structure and composition of the studied material, therefore this issue will be discussed in the following chapters.

1.2 Application of thin LiNbO₃ films

The idea of ferroelectric–semiconductor integration was realized in the 1960s. Since then a wide range of multifunctional devices utilizing the unique properties of ferroelectrics (nonvolatile memory, electro-optical modulation, ferroelectric sensitivity) have been created. It this section we will briefly review some of the most popular devices.

1.2.1 Waveguide-based optical devices

One of the potential applications of lithium niobate as an elemental basis of electro-optical devices is high speed (>20 GHz) modulators. Thin $LiNbO_3$ films contrary to bulk material, provide higher intensity per unit power in waveguides, and hence a stronger nonlinear optical effect and shorter interaction length. The ability to fabricate the films on various substrates extends the spectrum of their possible applications.

One of the key issues associated with the application of thin $LiNbO_3$ films in waveguides is the revelation of the mechanisms allowing exclusion or minimization of optical losses (absorption, leakage and bulk scattering, surface scattering and scattering triggered by grain boundaries scattering) and to satisfy the requirements of the film grain boundaries structure.

Optical leakage take place, for instance, when a thin film waveguide is fabricated onto a substrate with high refractive index (for example Si with $n = 3.8$ for $\lambda = 633$ nm) and an optical mode will inevitably 'leak' into the substrate. The degree of leakage depends on the film thickness. To achieve the maximum allowed leakage of 1 dB cm^{-1} formation of $LiNbO_3$ films with thickness about 400 nm is required [14].

Bulk scattering is an essential process for the optical mode coupling in a waveguide when the optical mode interacts with the radiative one. The change in the refractive index inevitably leads to scatter in the waveguide. This is caused by any inhomogeneities of the waveguide material: the presence of amorphous inclusions, parasitic phases ($LiNb_3O_8$, Li_3NbO_4), grain disordering in a polycrystalline film. Even in epitaxial films the significant bulk scattering can be observed due to the presence of sub-grains and mechanical strengths. The surface scattering and scattering at the grain boundaries depends greatly on surface roughness and can overwhelm the bulk scattering. Roughness of the order of 1 nm is critical, posing a complex technological problem, which requires a solution at the stage of $LiNbO_3$ film growth. It was shown [15] that regardless of the fabrication method of thin ferroelectric films onto sapphire substrates, there is a stable correlation between film thickness and surface roughness which was 1% of the thickness. The authors explained this tendency by the effect of grain boundary grooves formation due to competition between the surface energy of grain boundaries and the energy of the free surface in the process of island coalescence. Total energy minimization originated from the groove formation.

As a result, the main way to decrease the optical losses is fabrication of high quality two-axes textured or epitaxial $LiNbO_3$ films which is a challenging problem.

1.2.2 Surface acoustic wave devices

Surface acoustic waves (SAW) transfer the major part of their energy in the surface layer of a material within the thickness of about one or two acoustic wavelengths. This allows realizing their interaction with matter on such a surface. Moreover, SAW can have an electric field associated with them if they are generated in a piezoelectric material. This interaction leads to changes in the wave characteristics (amplitude, phase, speed etc) that underlie the operation of the chemical sensors. The operation principle (described in detail in some comprehensive reviews, for example in [16]) will not be discussed here, we will focus only on the requirements of a material for its fabrication.

Materials for SAW devices should possess primarily the following properties: piezoelectric effect, high Curie temperature, significant resistivity, high piezoresponse. Lithium niobate has all listed properties, so it is ranked as the most desired among piezoelectric materials such as $Pb(ZrTi)O_3$, $BaTiO_3$ and $Bi_4Ti_3O_{12}$, surpassing the competitors. Specifically, the extremely high Curie temperature and piezoelectric constant which characterize lithium niobate make it a unique material for high-temperature acoustic sensors.

It was emphasized [17–19], that the parameters of SAW filters are affected by the structure and surface morphology of $LiNbO_3$ to a great extent. It was demonstrated [18], that the SAW filters had poor characteristics due to high surface roughness and porosity of $LiNbO_3$ films. The authors of [19] associated the existence of preferable orientation and small surface roughness to the basic film characteristics for SAW filters that depend on fabrication technique. Emphasizing the challenge in the formation of c-oriented $LiNbO_3$ films in the CVD process, the authors of [20], demonstrated that the fabricated SAW device operated at a frequency of 6.3 GHz, which corresponds to a velocity of SAW of the order 12 500 m s^{-1}, i.e. exceeding the analogous characteristic for bulk LiNbO$_3$.

In [21] the fabricated SAW filter based on LiNbO$_3$ films, deposited by RFMS method onto diamond substrates was investigated. It reported the strong dependence of the structure on fabrication regimes and offered the optimal parameters (substrate temperature, RF magnetron power, reactive gas composition). Single phase films with low surface roughness ($\sim$10 nm) provided the operation of the SAW filters at the frequency of 2 GHz and SAW speed of 8200 m s^{-1}.

1.2.3 Memory units and neuromorphic systems

Thin LiNbO$_3$ films can be the basis for creation of perspective types of RAM[1], such as FRAM (FeRAM)[2] and RRAM (ReRAM)[3].

There have been a wide range of works focusing on the study of the possible application of thin LiNbO$_3$ films in the FRAM units (for example [22–24]). The following issues arise:

- fabrication of the single phase c-oriented and defect-free LiNbO$_3$ films with ferroelectric properties close to the single crystal lithium niobate;
- creating an ideal interface at the film/substrate heterostructure.

Provided these issues are solved, thin LiNbO$_3$ films as a functional element can replace the traditional ferroelectric materials (PbZr$_x$Ti$_{1-x}$O$_3$, BaTiO$_3$ and SrBi$_2$Ta$_2$O$_9$) in the contemporary FRAM units. The use of MOS-like heterostructures metal/LiNbO$_3$/Si potentially allows the introducion of FRAM units, applying two additional masking steps in the process of the fabrication of the regular CMOS[4]-structures. It makes possible the integration of FRAM to microcontrollers. Simplification of the technological stages of creation of the FRAM units will allow competition with currently dominating flash-memory due to their significant advantages [25].

The possibility of utilizing of LiNbO$_3$ films in RRAM units and neuromorphic systems was demonstrated in [26–29]. The main aim of this streamline is the development of the fundamental model describing the behavior of memristors. For example, it was reported [30, 31] on the memristive effect in TiO$_2$ films, where the change of resistance was associated with the motion of oxygen vacancies, activated by electrical current (the electroforming effect). In [32], the memristive effect was revealed in VO$_2$ films, which was caused by the insulator–metal transition in these structures.

The memristive effect in LiNbO$_3$-based heterostructures is explained in [26, 29] by the motion of oxygen vacancies, however, the resulting theoretical prediction does

---

[1] RAM—random-access memory.
[2] FRAM—ferroelectric random-access memory.
[3] RRAM—resistive random-access memory.
[4] CMOS—complementary metal-oxide-semiconductor.

not correlate with experimental data (see [27]). Moreover, some evidence, pointing to the role of Li ions in the process of memristive switching, were presented in [28].

Currently, $LiNbO_3$ seems to be an alternative material for RRAM, since the solutions based on this material are more technologically complex, so it is hard for it to compete with the RRAM technology on the basis of amorphous SiO_2. Nevertheless, $LiNbO_3$ has more potential in the context of neuromorphic systems [26, 28].

As a result of the creation of RAM units on the basis of FRAM and RRAM it will be possible to get more reliable, high speed, durable and protected storage memory compared to flash-memory units and HDD[5]. At the same time the neuromorphic systems could become the basis for creating the self-learning artificial intellect.

The potentially possible application of $LiNbO_3$ films is expanded greatly due to the revelation of ferromagnetism in them. It was proposed in [33, 34] to dope the films by ion implantation of the Co ions, when Co ions replace the Nb ions in lithium niobate. This mechanism is supported by the fact that oxygen vacancies are responsible for ferromagnetic properties in $LiNbO_3$ nanoparticles [35, 36]. As stated in [35], the mechanical–chemical synthesis of $LiNbO_3$ with subsequent annealing, leads to the formation of a high concentration of oxygen vacancy and allows the fabrication of samples manifesting ferromagnetic properties.

Although practical application of the multiferroic properties of lithium niobate is at the development stage, active and promising works on the $LiNbO_3$ films synthesis is a premise for their integration in optoelectronic devices.

1.3 Fabrication methods of thin $LiNbO_3$ films

It is important when choosing an appropriate synthesis method of thin $LiNbO_3$ films to lean on its possibility to preserve the elemental composition and structure of a bulk material, and also on the reproductive ability of this technique. Various fabrication methods have been proposed to satisfy these conditions, namely sol–gel, pulsed-laser deposition (PLD), discrete thermal evaporation in vacuum, radio frequency magnetron sputtering (RFMS), chemical vapor deposition (CVD), liquid phase epitaxy (LPE), ion-beam sputtering (IBS), Pechini method [37].

1.3.1 Liquid phase epitaxy

The possibility of the synthesis of $LiNbO_3$ films by LPE onto the surface of single crystal $LiTaO_3$ was demonstrated in 1975 [38, 39]. Although the the LPE method is currently rarely used for the fabrication $LiNbO_3$ films, some authors reported the successful fabrication of optical waveguides onto the $LiNbO_3$ single crystals [40, 41]. It was emphasized that when the films were doped by Zn atoms, the surface roughness increased with increasing the dopant concentration. At this doping the low degree of optical losses was achieved (0.154 ± 0.002 dB mm^{-1}, $\lambda = 1$ μm), and the surface roughness reached the lowest value (0.2–0.3 nm). In the process of

---

[5] HDD—hard (magnetic) disk drive.

pseudomorphic epitaxial growth the intermediate Zn:LiNbO$_3$ layer with the Zn content depending on substrate temperature is formed. Moreover, the refractive index difference at the film–substrate interface depends on the Zn concentration in LiNbO$_3$ film [40].

1.3.2 Chemical vapor deposition

The reactive gas mixture in a reactive chamber influences the synthesis of LiNbO$_3$ films in the CVD process. Polycrystalline films consisting of two phases, LiNbO$_3$ and LiNb$_3$O$_8$, are formed on the sapphire substrates at low reactive gas pressure (~266.6 Pa) [42]. The increase in reactive gas pressure up to ~2666 Pa leads to the formation of single phase LiNbO$_3$ films with <0001> texture. The value of mechanical stress declined from 1.3 GPa to 0.9 GPa with increase of pressure from 666.5 Pa to 101.3 kPa.

1.3.3 Sol–gel process

The first demonstration of the possibility of fabrication of complex oxides was in [43], and the sol–gel method has been widely used for deposition of LiNbO$_3$ films since the 1980s [44].

Due to the high degree of gel reactivity, film formation is fulfilled at significantly lower temperatures compared to other methods. The sol–gel method is extremely time- and resource-consuming (expensive precursors are used for fabrication of a gel) and the synthesized LiNbO$_3$ films normally consist of extra non-ferroelectric phases (LiNb$_3$O$_8$ and Nb$_2$O$_5$).

The formation of the LiNb$_3$O$_8$ phase can be associated with the phase reaction between LiNbO$_3$ and the natural oxide existing on a Si substrate [45]. The amount of LiNb$_3$O$_8$ phase in synthesized films is affected by the annealing temperature.

Surface roughness depends on the deposition conditions as well as on molar concentration of compounds in a solution and changes from 10.6 nm to 16 nm when concentration rises from 0.25 to 1 mol l^{-1} [46].

1.3.4 Pulsed-laser deposition

The possibility of deposition of ferroelectric films by means of the PLD method was first demonstrated in [47]. For synthesis of LiNbO$_3$ films this method became widely used in the 1990s [48–50].

The PLD method is based on the process of ablation of a target under the influence of short (20–30 s) pulses of the focused laser radiation. The substrate holder is located in the atmosphere with high oxygen pressure, required for oxide formation. Film parameters depend on many factors: distribution of compounds during the ablation, reactive gas pressure, substrate–target distance, energy and frequency of laser radiation.

LiNbO$_3$ films, deposited by the PLD technique, manifest non-uniform composition and structure. Early works [48, 49] revealed the formation of a Li-deficit phase LiNb$_3$O$_8$. It was proposed that Li-rich be used as targets as an effective method to avoid this issue [48]. Also, there is evidence of a relationship between the appearance of $LiNb_3O_8$ phase and the relative position of a substrate and a target as well as the presence of oxygen in a reactive chamber [51].

Results of [52–54] demonstrated the dependence of the elemental composition, structure and surface morphology on the target structure and PLD regimes. Uniform single phase and crack-free $LiNbO_3$ films with grain size of 50–200 nm were deposited by the PLD method onto sapphire substrates and SiO_2/Si heterostructures at the substrate temperature $T_{sub} = 700$ °C [52]. Interestingly the films deposited onto sapphire demonstrated more transparency than films on SiO_2/Si heterostructures.

In [53] the nonlinear dependence between the average roughness and partial oxygen pressure for $LiNbO_3$ films deposited onto diamond/Si heterostructures ($T_{sub} = 650$ °C) is revealed. This dependence was described in terms of competition between two processes: scattering by the atoms of a reactive environment (at high pressure) and re-sputtering of the growth front (at low pressure). The successful synthesis of highly oriented $LiNbO_3$ films with <0001> texture onto SiO_2/Si substrates at $T_{sub} = 650$ °C was also reported [54]. The following dependence of degree of preferable orientation on the substrate temperature in the process of film formation was stated: increase of T_{sub} leads to a decrease in the degree of orientation up to the formation of films with arbitrary grain orientation at 750 °C. It is important to stress, that single phase c-oriented $LiNbO_3$ films are formed only at the oxygen pressure of 40 Pa in a reactive chamber and further increase in the oxygen pressure suppresses the formation of $LiNb_3O_8$ phase.

The minimal roughness (4.3 nm) was achieved when the substrate temperature was 600 °C; at $T_{sub} = 500$ °C it rose to 5.1 nm and at $T_{sub} = 700$ °C roughness increased to 5.8 nm. Optic losses minimum corresponded to the minimal roughness. The authors explained this dependence of roughness on T_{sub} through the change of mobility of adatoms.

1.3.5 Radio frequency magnetron sputtering

The first time $LiNbO_3$ films were fabricated in the sputtering triode configuration from the powdered $LiNbO_3$ target was in [55]. Since then, $LiNbO_3$ films had been synthesized by radio frequency diode sputtering (rf-sputtering) [56–59], which is associated with low deposition rate and composition inhomogeneity of deposited films.

With the development of magnetron sputtering and radio frequency magnetron sputtering (RFMS), rf-sputtering was totally replaced by RFMS having the following advantages:
- lower reactive gas pressure, providing the appropriate free path length for ion motion in the space charge area;
- ability to regulate the ion energy in a wide range through magnetic field at the constant source power;
- high rate of sputtering;
- independence of the sputtering coefficients of a material on its melting point;
- reproducibility of elemental composition of a sputtered material.

Initially, the synthesis of $LiNbO_3$ films by the RFMS method was attempted in the second half of the 1980s [60], but the study of the influence of technological parameters and plasma effect (the ion assisting effect) on the structure of deposited $LiNbO_3$ films is ongoing.

Table 1.1 presents advantages and disadvantages corresponding to each method discussed above.

As shown in table 1.1 the RFMS process is one of the most effective deposition methods of $LiNbO_3$ films, satisfying the basic requirements, dictated by the applications.

1.4 Fundamentals of RFMS method, an ion-beam sputtering method and their critical parameters

Sputtering can be defined as an ejection of atoms or molecules from the target material due to its bombardment by high energetic ions of reactive plasma.

If *the direct* voltage, not exceeding the breakdown voltage, is applied to two electrodes placed in a gas, the current between these electrodes is infinitesimal. Traditionally, argon is used as a reactive gas, which is ionized by electrons between the electrode conjugated with a target (cathode) and substrate (anode). Due to emission from the cathode the current appears and it vanishes when the emission stops. This current rises as the distance between electrodes increases, since electrons experience more collisions ionizing reactive gas atoms in the chamber. Every collision generates an electron as well as an ion produced through ionization and accelerating by field. At relatively high voltage some ions reach and bombard the target surface, inducing sputtering of its surface and the formation of secondary electrons. When ion current plays a significant role in total current, the ions start concentrating near the cathode, producing the localized space charge and increasing electric field. When the breakdown is induced, the number of secondary electrons is enough to maintain autonomous plasma discharge. Electrons due to their low mass, travel through the space charge region very fast and reach the neutral region, where the number of electrons and ions is nearly equal. This area is called plasma. Because the ions are shielded by electrons, they 'do not know' about the existence of the electrode and they move via diffusion. As soon as an ion reaches the space charge area, it starts 'feeling' the potential at the cathode and the ion rushes to it. Atoms, ejected from the target due to bombardment, are able to deposit onto a substrate placed in the reactive chamber, which leads to steady growth of a thin film on its surface. The process described above underlies the basic approach of ion sputtering of materials (IBS method).

One of the main disadvantages is the relatively low film deposition rate and high cathode voltage.

To increase efficiency of ionization of a reactive gas by electrons, to decrease reactive gas pressure and to reduce the cathode voltage, the RFMS technique was proposed.

Table 1.1. Comparative analysis of advantages and disadvantages associated with the most popular deposition techniques.

Method	Advantages	Disadvantages
Pulsed lased deposition (PLD)	• Elemental composition is conserved. • Deposition at relatively high oxygen pressure. • Relatively high deposition rate.	• Small size of the laser impact area leads to non-uniform sputtering. • Deviation from elemental composition of a target due to long lasting ablation. • Careful selection of many technological parameters (many of them are dependent on each other). • The presence of the drops in a flow, that inevitably makes fabrication of layered structures difficult. • Relatively high process temperature.
Sol–gel process	• Homogeneity of deposited films. • It is possible to deposit films onto wide substrate areas. • Simplicity of integration with existing semiconductor technology.	• Structure and electrical properties of deposited films are influenced by the process conditions. • High probability of crack of the films due to solvent evaporation. • Particle agglomeration and non-homogeneous composition. • Subsequent annealing is required.
Chemical vapor deposition-process (CVD)	• Capability of film synthesis onto the stepped surfaces. • Deposition at relatively high oxygen pressure. • High deposition rate. • It is possible to deposit high purity films onto wide substrate areas.	• Careful selection of a precursor (usually extremely toxic and corrosive) is required. • Relatively high process temperature (up to 900 °C). • Complexity of the process and strong dependence of the films properties on regimes.
Liquid phase epitaxy (LPE)	• The simplicity of equipment. • Capability of the films doping. • Low compound toxicity.	• Inhomogeneity of the film structure. • High surface roughness. • The presence of the solvent compounds and a catalyst in deposited films. • Low reproductivity of stoichiometry and structural characteristics.
RF-magnetron sputtering (RFMS)	• Controllable sputtering process. • Suitable for synthesis in a wide range of thin film materials: metals, semiconductors, dielectrics and complex oxides. • Relatively high deposition rate. • Capability of correspondence of elemental composition of the film and target. • It is possible to deposit thin films onto wide substrate areas.	• Target erosion causing its degradation. • Complexity of the plasma effect on growing process.

RFMS offers the following advantages compared to traditional diode sputtering systems without magnetic field:

- Lower reactive gas pressure, providing collision-free motion of ions in the space charge area.
- No diffusion of the sputtered particles.
- Decrease of the ion energy up to hundreds of electron-volts and the ability to regulate this energy in a wide range via magnetic field at the constant discharge power. It allows the setting of the optimal energy of bombarding ions required for a given process.
- Sputtering rate increase compared to sputtering systems without magnetic field at the same power due to more efficient use of energetic electrons in plasma.
- Nearly zero dependence of sputtering coefficients on the melting point of materials.
- High composition reproducibility of a sputtering material.
- Ability to sputter dielectric materials.

The magnetron sputtering system is placed in a reactive vacuum chamber, where the pressure does not exceed 10^{-4} Pa before supplying the reactive gas and it rises up to 10^{-1} Pa when the gas is added.

Figure 1.3(a) demonstrates the basic scheme of RFMS system and figure 1.3(b) shows one of the schemes of a ring magnetron used in RFMS systems.

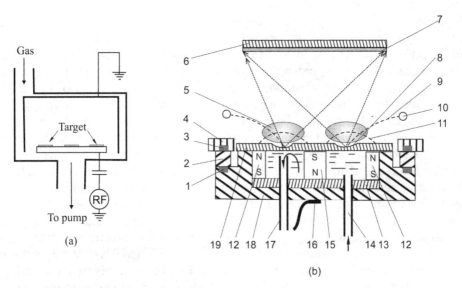

Figure 1.3. Schematic diagram of RFMS system (a) and the schemes of a ring magnetron used in RFMS systems (b). 1,3—vacuum gaskets, 2—isolating ring, 4—chamber flange, 5,8—plasma zone and erosion zone, 6—substrate, 7—thin film, 9,11—electric and magnetic fields, 10—anode, 12,15—peripheral and central magnets, 13—base of the magnetic block, 14,17—pipes, 16—clamp, 18—hull, 19—target.

In the ring planar magnetron (figure 1.3(b)) all components are mounted in a hull (18), attached to the reactive chamber through the intermediate isolating ring (2) and a flange (4) with the vacuum gaskets (1) and (3) (see figure 1.3(b)). Disc-shape target-cathode (19) is cooled by water flowing through pipes (14) and (17). Voltage is supplied via the clamp (16). Under the cathode the magnet block, consisting of the central magnet (15) and the peripheral magnet (12) is located. Magnets are mounted on the base of the block (13). The magnet block generates a magnetic field and its component is parallel to the cathode plane. Anode (10) is situated above the cathode and is grounded, ensuring generation of electric field (9) with its component perpendicular to the cathode plane.

Physically the magnet system affects the plasma discharge properties. The magnetic field, affecting the glowing discharge, primarily changes the character of the motion of electrons rather than ions having relatively high mass. Magnetic field lines are closed between the system poles.

The surface of a target, located at the exits and entrances of the magnetic field lines, is sputtered more intensively and has the shape of the closed groove. For magnetrons, shown in figure 1.3(b), this groove is a ring. From this point of view the magnetic field effect is equivalent to an increase in reactive gas pressure. Another important effect is the 'magnetic trap effect' decreasing greatly the radial out-of diffusion of electrons from the discharge zone, eliminating the loss of electrons.

One of the key characteristics of the RFMS process is the plasma potential—the electrical potential of the central part of the glowing discharge measured from the grounded electrode potential. When a high frequency signal is applied to the cathode, the plasma potential follows the applied signal frequency. Averaged over time the plasma potential changes from the potential associated with the ground electrode to a few hundred volts depending on magnetron power (RF signal), reactive gas pressure and the electrode geometry. The asymmetric electrode configuration, usually used in RFMS systems, provides greater reactor efficiency. Electrons as the carriers, having significantly higher mobility compared to ions, follow the RF voltage on a cathode which normally has a frequency of 13.5 MHz, whereas ions do not follow. As a result, ions follow only the voltage averaged over time on the cathode, inducing the intrinsic plasma field. This field generates the ion flux towards the target and causes its bombardment. The intensity of this bombardment depends on magnetron power, reactive gas pressure, target composition, the intrinsic field and other parameters. Apart from argon, gaseous oxygen is added to the reactive chamber to provide the formation of the oxide compounds with the desired composition on a substrate. It is important to emphasize, that the relative position of a target and a substrate is one of the key parameters affecting film formation. For instance, when the substrate is situated exactly over the target erosion zone at a short distance, the ions of plasma, reflected from the target during its sputtering, influence the surface of a growing film. Further, this regime we call *the ion assist regime*. The degree of this effect depends on magnetron power, gas pressure in the reactive chamber, target–substrate distance and their relative orientation.

Because sputtering is a physical rather than chemical process, it allows a wide range of thin film materials to grow on various substrates. Nevertheless, the sputtering

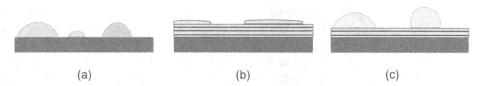

Figure 1.4. The schematic representation of three basic models of the epitaxial film growth. The Volmer–Weber (island) (a), the Frank and van der Merwe (layer) model (b), the Stranski–Krastanov (layer followed by island) model (c).

process can vary depending on target composition and some other parameters it is not guaranteed that the grown films will have compositions exactly corresponding to the material of the target. Atoms of a target, depositing onto a substrate surface, are not simply fixed at a certain position. They can either diffuse (the Gibbs–Thomson effect) or move along the surface provided the absence of secondary evaporation from the surface. Where the atoms move along the surface they can combine with other atoms, forming clusters of the target material (nucleation). Further evolution of this process can be realized via several possible scenarios [61], leading to the formation of thin $LiNbO_3$ film (figure 1.4) provided the sputtering parameters are carefully controlled. Because various materials have different adhesion energy depending on a substrate–film combination, the epitaxial growth can be classified in the frameworks of three basic models: Volmer–Weber (island), Frank and van der Merwe (layer) and Stranski–Krastanov (layer followed by island).

According to the Volmer–Weber model (figure 1.4(a)), at the beginning, the aggregates of nucleations of a new phase are formed on a substrate, which form a continuous layer due to coalescence at the later stages. This mechanism takes place in the case of poor adhesion of crystallites and the surface. The Frank and van der Merwe mechanism (figure 1.4(b)) is implemented by the strong adhesion between crystallites and a substrate, causing the layered film formation (layer by layer). The Stranski–Krastanov model (figure 1.4(c)) reflects an intermediate case between these two approaches and is usually realized in metal–metal and metal–semiconductor systems. We suggest that for the deposition methods and regimes described in this book the Volmer–Weber model is used, because many investigators argue that this model describes more correctly the synthesis of $LiNbO_3$ films by sputtering in vacuum. For instance, the authors of [62] deposited c-oriented $LiNbO_3$ films onto $Si–SiO_2$ heterostructures by the PLD method. They reported the formation of 3D islands (nucleations) at the beginning stages of deposition, testifying to the Volmer–Weber model. Another group of investigators also reported the nucleation of islands with sizes of around 105 nm at the earlier deposition stages [63, 64]. Further deposition leads to inducing coalescence with formation of thin $LiNbO_3$ film.

The most critical parameters of RF-magnetron sputtering method are as follows:
- Type and temperature of a substrate.
- Reactive gas composition.
- Reactive gas pressure.
- Reflected magnetron power.
- Composition of a target and its position relative to the substrate.

Therefore, a lot of effort has been focused on revealing the dependence between technological parameters of RFMS listed above and structure, composition and electrical properties of fabricated $LiNbO_3$ films.

Deviation of *the target material* from stoichiometry, the target density and the presence of impurities greatly influence the degree of crystallinity of synthesized films. Specifically, in [65], the properties of films, fabricated by sputtering of a single crystal target, grown by the Czochralski method and the target fabricated by sintering of Li_2CO_3 and Nb_2O_5 powders at temperature of 1200 °C according to the following reaction were compared:

$$Li_2CO_3 + Nb_2O_5 \rightarrow 2LiNbO_3 + CO_2\uparrow \qquad (1.1)$$

Study of the composition revealed that Li_2CO_3 oxide is not removed entirely from the target, forming a surface layer with preferable orientation along $LiNbO_3$ polar axes due to segregation. Bombardment of such a target in the RFMS process leads inevitably to bringing the Li_2CO_3 oxide into the synthesized film, causing the formation of Li-poor $LiNbO_3$ layers. Another undesirable effect is that Li_2CO_3 presented and dissolved in a film, generates point defects, inhibiting the formation of a large-block $LiNbO_3$ polycrystalline structure. Some investigators proposed [66] adding about 5% of Li_2O oxide to the target composition to ensure the formation of stoichiometric $LiNbO_3$ films. Another research group used the Li-rich target provided optimal elemental composition in a reactive chamber through evaporation of lithium [67].

Evidently, the use of a single crystal $LiNbO_3$ target is more favorable for preserving the initial elemental composition and formation of films with a high degree of crystallinity.

RF-power influences greatly the properties of $LiNbO_3$ films. Figure 1.5 shows an XRD pattern of $LiNbO_3$ films, fabricated under different RF-power at the substrate temperature of 580 °C [67].

As seen from figure 1.5, RF-power affects film composition and the degree of crystallinity. The fact that only amorphous films are formed at RF-power of 70 W is explained in literature: this power is not sufficient for the stable nucleation on the substrate surface. On the other hand it is pointed out that at RF-power of 100 W polycrystalline films with (012) preferable orientation are formed, due to minimal surface energy associated with $LiNbO_3$ in this plane [67]. A further increase of RF-power leads to the formation of extra phases of $LiNb_3O_8$ and $Li_{1.9}Nb_2O_5$ and the disappearance of oriented growth of a film. Furthermore, the authors of [67, 68] reported a nearly linear dependence of the film growth rate on RF-power, which was from 17 Å min^{-1} to 26 Å min^{-1} at RF power of 100 W. Other investigators also recommend that sputtering is conducted at RF-power of 100 W allowing the formation of highly oriented (006) $LiNbO_3$ films [68–70]. As pointed out in [69], sputtering at RF-power of 100 W corresponds to the formation of films with minimal surface roughness (see figure 1.6).

At the same time the degree of crystallinity rises with RF-power as can be seen in figure 1.7 through the increase in the intensity of the XRD peaks.

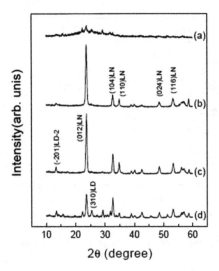

Figure 1.5. XRD patterns of the thin films deposited at the substrate temperature of 580 °C and 1 mTorr working pressure. The RF-power is 70 W (a), 100 W (b), 130 W (c), 170 W (d); LN, LD and LD-2 stand for $LiNbO_3$, $LiNb_3O_8$ and $Li_{1.9}Nb_2O_5$, respectively [67]. Copyright (1999) The Japan Society of Applied Physics.

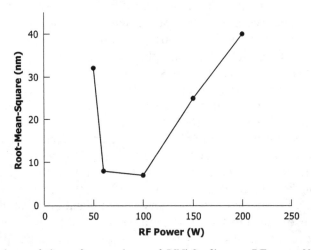

Figure 1.6. Dependence of the surface roughness of $LiNbO_3$ films on RF-power [69]. Reprinted with permission from Elsevier. Copyright 2007.

A detailed investigation [70] showed that relatively high RF-power may lead to extensive heating of a target, inducing not only its cracking, but also dissociation of lithium and oxygen and consequently depletion of the target by lithium. As a consequence, the fabricated films consist of a Li-poor non-ferroelectric $LiNb_3O_8$ phase. RF-power of 100 W is an optimal parameter for synthesis of $LiNbO_3$ films with the index of refraction very close to a bulk material [58].

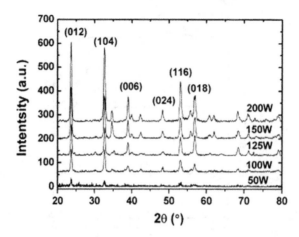

Figure 1.7. XRD pattern of LiNbO₃ films, fabricated by RFMS as a function of RF-power [69]. Reprinted with permission from Elsevier. Copyright 2007.

The *temperature of a substrate* and its type in the RFMS process directly affect the growth rate, structure and composition of LiNbO₃ films. Whereas the influence of the substrate temperature on the growth rate is not so pronounced, the degree of crystallinity and film orientation are influenced by this parameter to a great extent [67, 69]. There is evidence [67], that at the reactive gas pressure of 1 mTorr and RF-power only amorphous LiNbO₃ films are formed on non-heated substrates, and a substrate temperature of 560 °C is optimal for oriented growth of the films.

At the excessive substrate temperatures, evaporation of lithium from the surface of a deposited film occurs, leading to the formation of an undesired Li-poor LiNb₃O₈ phase [71]. As the type of substrate is a concern, it is generally accepted that the ideal choice corresponds to the substrate with lattice parameters close to LiNbO₃, to decrease surface strain in a film and to eliminate defect concentration at the film–substrate interface. Various intermediate layers such as SiO₂, Si₃N₄/SiO₂, and ZnO are used for this purpose. For instance, some investigators fabricated a high oriented ZnO film (an intermediate layer with thickness of 100 nm) onto the silicon substrate, allowing growth of c-oriented LiNbO₃ films with minimal lattice strain [70].

Perhaps, one of the most critical parameters of RFMS is *the reactive gas pressure and composition*. As was mentioned above, reactive gas pressure greatly influences properties of plasma, as well as the growth rate, degree of orientation, surface roughness of synthesized films. Some researchers stressed, that reactive gas pressure directly affects the concentration of Li in a film [67]. The origin of this is that at high gas pressure the mean free path of Li ions, having relatively low atomic mass, declines, leading to their strong scattering in plasma. As a result, the number of ions reaching a substrate drops, which leads to the formation of the Li-poor LiNb₃O₈ phase.

The authors of [71, 72] stressed that films, deposited at low pressure in the reactive chamber (1–2 mTorr), manifested a high degree of crystallinity and the preferred orientation (012) [71] and (006) [72]. They recommend performing sputtering at the

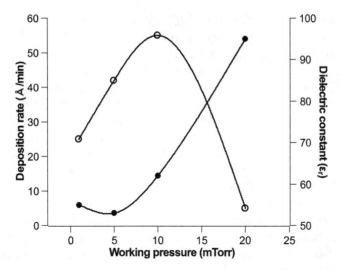

Figure 1.8. Deposition rate and dielectric constant as a function of working pressure [71]. Reprinted with permission from Elsevier. Copyright 1999.

RF-power and substrate temperature of 100 W and 600 °C, respectively. As regards dependence of the growth rate on gas pressure, it obeys a nonlinear dependence (see figure 1.8).

Initially, the increase of deposition rate occurs with working pressure, but then a sharp decrease takes place, which can be explained by the decrease in the mean free path of deposited compounds in plasma.

As mentioned earlier, because thin LiNbO₃ films, fabricated by various methods (and RFMS is not an exception) demonstrate Li deficit, the most natural way of managing this process is the careful selection of a reactive gas composition. Specifically, it was demonstrated in [73, 74], that the presence of oxygen as a reactive gas component, increases the partial pressure of lithium in the reactive chamber. On the other hand the use of pure oxygen as a reactive environment does not allow the fabrication of single phase LiNbO₃ films [49]. In this regard many efforts were focused on the search for an optimal Ar/O₂ ratio in the reactive chamber taking into account other parameters of RFMS and the desired properties of synthesized films. Based on the study of the influence of the presence of oxygen on the structure of LiNbO₃ films, deposited onto silicon substrates at a temperature of 575 °C and working pressure of 100 W, the authors of [69] recommended the use of the ratio Ar/O₂ = 1. In the recommended technological regimes high oriented (006) LiNbO₃ films with good optical properties are formed. On the other hand, it was demonstrated [70], that the concentration of oxygen in the reactive chamber has a profound effect on the lattice constant along the symmetry axis of synthesized c-oriented LiNbO₃ films and consequently on their surface tension. Based on detailed study, the authors of this work revealed that only in an Ar + O₂ plasma with ratio of components 80% + 20% are films with minimal strain and lattice constant close to bulk material ($c = 1.3867$ nm) formed. The authors concluded that the increase in concentration of oxygen in the reactive chamber leads to intensifying the bombardment of a deposited film and consequently, encouraging the insertion of oxygen in interstices of the crystal lattice causing its deformation. With that, it was shown in [75], that the films, fabricated at low oxygen pressure, manifest a lower degree of orientation and consist of $LiNb_3O_8$ which is accounted for by evaporation of lithium from the film surface at low pressure in the chamber. With the increase in pressure up to 30 Pa only single phase highly oriented (006) $LiNbO_3$ films are formed, but this effect becomes weaker with further pressure increase. Dependence of preferable orientation and degree of crystallinity on oxygen pressure was investigated in [53] in the framework of plasma dynamics. It was noted, that the kinetic energy of the plasma components and the size of a plasma area vary depending on oxygen pressure due to collisions between atoms and the substrate–target distance which provides optimal conditions for the synthesis of high quality films. With that, the kinetic energy of particles reaching a substrate should be restricted by the optimal value. At high oxygen pressure, the probability of collisions with plasma components increases, with decreasing the energy of bombarding ions that has a negative effect on the degree of crystallinity of a deposited film. On the other hand, due to the low intensity of collisions at low oxygen pressure, the bombarding ions reach the substrate with excessive energy generating a lot of defects in the film and preventing its oriented growth. Moreover, due to numerous collisions at high reactive gas pressure, Li atoms do not reach the substrate surface with sufficient amounts, leading to the formation of Li-poor phase ($LiNb_3O_8$). Nevertheless, many authors recommend using the Ar/O_2 reactive plasma with proportions of 60/40 to fabricate single phase highly oriented $LiNbO_3$ films [22, 71, 72, 76]. Specifically, it was demonstrated in [22], that in the case of an inappropriate choice of the Ar/O_2 ratio, as a rule, dissociation of $LiNbO_3$ takes place with formation of the Li-poor phases ($LiNb_3O_8$) and oxides like Nb_2O_5. There are three locations where it can potentially occur: at the target surface, in the plasma and in the film after its deposition. It was stressed, that the substrate temperature should not exceed 600 °C to prevent dissociation. As a result of research, aiming to optimize technological regimes, the authors concluded that the Ar/O_2 ratio of 65/35 is required for the deposition of highly oriented (006) $LiNbO_3$ films. As regards the Nb_2O_5 oxide, it is the product of dissociation of the $LiNb_3O_8$ phase according to the following reactions [22]:

$$3LiNbO_3 \rightarrow LiNb_3O_8 + Li_2O$$
$$2LiNb_3O_8 \rightarrow 3Nb_2O_5 + Li_2O \qquad (1.2)$$

Nevertheless, it is important to stress that the Ar/O_2 ratio is not an independent parameter, rather it should be considered coupled with other parameters influencing the properties of reactive plasma such as working pressure and RF-power.

Apart from issues discussed above, it is important to mention the methods of post-deposition treatment of $LiNbO_3$ films. One of the most popular methods to increase the degree of crystallinity of deposited films, and decrease defect concentration and mechanical strain is thermal annealing (TA) in a vacuum or in a gas environment. Pure oxygen or air is the most frequently used gas environment for TA. Oxygen atmosphere affects the crystallization process of $LiNbO_3$ films [77, 78]. The role of oxygen is to prevent evaporation of light compounds such as Li_2O, to control the stoichiometry and to improve electrical properties. The majority of authors reported the formation of amorphous $LiNbO_3$ films on non-heated substrates [70, 77, 79], which manifest ferroelectric properties after TA. The surface roughness depends on the annealing temperature nonlinearly. For example, it was shown in [77], that the minimal roughness of 2.6 nm corresponding to amorphous films increased up to 210 nm for the films, annealed at 550 °C, and then it declined up to 80 nm associated with the annealing temperature of 1000 °C. The minimal average surface roughness (15 nm) corresponded to the annealing temperature of 700 °C. Thermal annealing of amorphous films led to the formation of films with random orientation of grains [77, 80]. When oriented films deposited onto heated substrates had undergone thermal annealing, the final effect depended on the temperature and duration of TA. After TA of $LiNbO_3$ films in air the O/Nb ratio approaches the stoichiometric one [81]. Studying the TA of $LiNbO_3$ films in a static and a dynamic oxygen atmosphere showed that films annealed in the dynamic regime demonstrated more uniform size distribution of polycrystalline grains compared to films annealed in the static atmosphere [82]. The authors explained this phenomenon by the fact that during oxygen flow more intensive evaporation of components occurs from the film surface, causing the decline in its growth and consequently lower surface roughness. Nevertheless, most researchers report the formation of a Li-poor (parasitic) $LiNb_3O_8$ phase after TA of as-grown $LiNbO_3$ films in both vacuum and oxygen atmosphere [78, 83, 84].

Detailed study of the structural properties of $LiNbO_3$ films after TA resulted in an original model, which describes the formation of $LiNb_3O_8$ phase in the process of thermal annealing. According to this model, the following reactions take place simultaneously during TA:

$$c-LiNbO_3 \rightarrow c-LiNb_3O_8 + Li_2O \uparrow ,$$
$$c-LiNbO_3 \rightarrow c-LiNbO_{3-x} + (x/2)O_2 \uparrow . \tag{1.3}$$

The first equation in (1.3) corresponds to the loss of Li and this process prevails at the beginning stage, whereas the second equation describes the loss of oxygen. The hexagonal atom position in the (0001) crystallographic plane of $LiNbO_3$ is comparable with their position in the $(\bar{6}02)$ plane of $LiNb_3O_8$ with slight lattice mismatch. It is most likely that $LiNb_3O_8$ crystallites precipitate while maintaining their epitaxial relationship with the original $LiNbO_3$ crystallites. Figure 1.9 clarifies the mechanism of reactions and the structural phase conversion taking place during TA of $LiNbO_3$ films.

At the early stages of annealing in *a vacuum*, Li_2O is desorbed from the surface and grain boundaries, whereas the single grains are merged, forming interphase boundaries (figure 1.9(a)). When the desorption of Li_2O slows down (saturation),

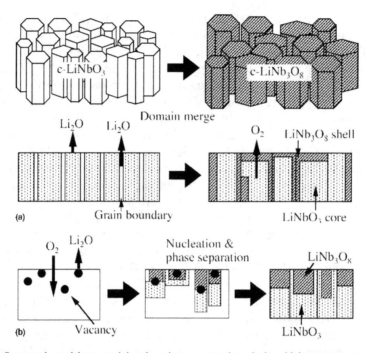

Figure 1.9. Structural model to explain the phase conversion during high-temperature annealing of (a) c-oriented LiNbO₃ textured film in a vacuum (upper and middle rows) and (b) a-LiNbO₃ film in O₂ ambient (lower row). The upper row is the nest view, and the middle and bottom rows are cross-sectional views. Reproduced from [83] with permission.

oxygen atoms continue evaporating from the inner areas of a film. Finally, the textured crystallites of LiNbO₃ are embedded in the crystalline 'shell' of LiNb₃O₈ phase, forming the structure in which each grain has a core (LiNbO₃) and an outer layer (LiNb₃O₈).

During TA in an O₂ atmosphere the oxygen molecules penetrate a film, preventing the formation of vacancies that block the migration ability of adatoms. In the framework of this model, during the incubation period the Li₂O evaporates according the following reaction [83]:

$$a-\text{LiNbO}_3 \rightarrow a-\text{Li}_{1-2x}\text{NbO}_{3-x} + x \cdot \text{Li}_2\text{O}\uparrow \qquad (1.4)$$

This process creates vacancies, facilitating the atomic motion and causing the nucleation of LiNbO₃–LiNb₃O₈ pairs due to the phase separation as follows:

$$a-\text{Li}_{1-2x}\text{NbO}_{3-x} \rightarrow (1-3x) \cdot c-\text{LiNbO}_3 + x \cdot c-\text{LiNb}_3\text{O}_8 \qquad (1.5)$$

When the nuclei of LiNbO₃ and LiNb₃O₈ are formed, the processes of crystallization and phase transition continues with speed depending on the annealing conditions. It is worth noting that similar transitions are possible not only for crystalline but also for amorphous films.

1.5 Electrical properties and charge transport phenomena in LiNbO$_3$-based heterostructures

The fact that structural properties of LiNbO$_3$ films strongly depend on deposition parameters is widely reflected in many papers. However, the electrical properties, such as ferroelectric properties or charge transport in LiNbO$_3$-based heterostructures, are not presented systematically, rather they are reported as complementary parts of the structural study. Surprisingly, papers on the electrical properties of LiNbO$_3$ often report only the general tendency in current–voltage characteristics, dc or ac conductivity rather than quantitative parameters that can be derived from the observed electrical phenomena. In addition, the information regarding conduction mechanisms and the type of charge carriers is sometimes controversial. Below, based on literature sources, we present a systematic overview of charge transport phenomena in LiNbO$_3$-based heterostructures.

1.5.1 Charge carriers in LiNbO$_3$

Traditionally, charge carriers in single crystal LiNbO$_3$ are triggered by the point intrinsic or extrinsic defects [85, 86].

Three main sources of ionic defects in LiNbO$_3$ can be classified as follows: the impurity content, non-stoichiometry in Li/Nb and intrinsic ionic disorder. The intrinsic ionic disorder is a predominant source of the ionic defects. The shortage of Li during the growing process can be described by various possible processes of Li$_2$O out diffusion [85]:

$$\langle\langle LiNbO_3 \rangle\rangle \rightarrow Li_2O + 2V_{Li}' + V_O^{2\cdot}$$

$$\langle\langle LiNbO_3 \rangle\rangle \rightarrow 3Li_2O + 4V_{Li}' + Nb_{Li}^{4\cdot} \tag{1.6}$$

$$\langle\langle LiNbO_3 \rangle\rangle \rightarrow 3Li_2O + 4V_{Nb}' + 5Nb_{Li}^{4\cdot}$$

In these reactions V_{Li}, V_{Nb} and V_O are lithium, niobium and oxygen vacancies respectively, Nb$_{Li}$ is an Nb antisite defect at V_{Li} (the bullet and the prime represent the positive and negative charge states of a defect with respect to the host, respectively). The second reaction in equation (1.6) describes the most popular model. However, the generation of antisite defects and degree of filling the Li sites by Nb ions in general depends on the growth conditions [87, 88].

The variety of experimental results and different types of observed charge carriers are influenced by a wide range of growing conditions and fabrication methods of LiNbO$_3$ films.

The polaronic model, describing charge phenomena in lithium niobate, is one of the most widely used approaches [89]. A small polaron can be created when sufficiently strong interaction of a carrier with the surrounding lattice induces a self-trapping of the carrier at one site in condensed matter. Due to the vibration of the surrounding ions, the equilibrium positions of this carrier are displaced. Even insignificant irregularities of the lattice cause complete localization at one site. The main charge transport mechanisms in this situation are the thermally activated hopping or tunneling. The small bound polaron is created if electrons are captured at the $Nb_{Li}^{4\cdot}$ antisite defect (induced according to equation (1.6) by the concomitant lattice distortion). Furthermore, a bound bipolaron is formed when the nearest defect pair $Nb_{Li}^{4\cdot} - Nb_{Li}^{4\cdot}$ is capable of capturing two electrons. Additionally, the extrinsic ions, inevitably presented in $LiNbO_3$ (for instance Fe_{Li}^{2+}), are capable of capturing an electron, generating small bound electron polarons. The polaronic model in $LiNbO_3$, underlying many nonlinear phenomena in this oxide is supported by optical investigations [90–92].

Alternatively, experiments suggest [92, 93], that the hole polarons are the major charge carriers in $LiNbO_3$. In lithium niobate like other oxide materials the valence band holes can be trapped at negatively charged acceptor defects. Despite the fact that the oxygen ions in a lattice are equivalent, the hole–lattice coupling causes a hole localization, forming an isolated O^- ion. Recall, non-stoichiometric Li deficit leads to formation of intrinsic acceptors (Li vacancies V_{Li}) that can capture an isolated O^- ion generating the 'O^- next to V_{Li}' structure (see figure 1.10) [93].

Nonetheless, the electrical measurements suggest that electrons are the major carriers in $LiNbO_3$ films [94], because potentially free electrons can be generated in lithium niobate according to the following reactions [87]:

$$V_O \leftrightarrow V_O^{\cdot} + e^-$$

$$3O_O + 2V_{Li}' + Nb_{Nb} \leftrightarrow \frac{3}{2}O_2 + Nb_{Li}^{4\cdot} + 6e^- \qquad (1.7)$$

It is important to note, that primarily, regardless of the fabrication methods, the $LiNbO_3$ films are polycrystalline, so quite a high density of interface states (defects)

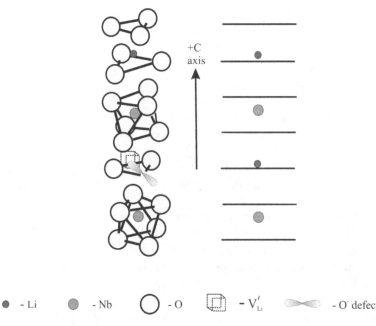

$\bullet$ - Li $\bullet$ - Nb $\bigcirc$ - O   ⬚ - V_{Li}' ∞ - O^- defect

Figure 1.10. Model of an 'O^- next to V_{Li}' structure in $LiNbO_3$.

are presented at the grain boundaries and act as donor centers. Regardless the type of major charge carriers, polycrystalline structure of fabricated $LiNbO_3$ films leads to additional complications in the interpretation of experimental results dealing with charge transport phenomena. In the next section we present a brief overview of the main charge transport mechanisms observed in $LiNbO_3$-based heterostructures.

Generally, all conduction mechanisms in an insulator can be divided into two main groups: bulk-limited (influenced by the properties of the dielectric itself) and contact-limited (affected by contact properties) [95]. However, two or more concurrent mechanisms can take place simultaneously, depending on experimental conditions (applied voltage, temperature, illumination, film thickness etc). Because electrical conductivity is influenced by temperature and applied voltage in different ways it is possible to isolate the charge transport mechanism.

1.5.2 Electrical conductivity limited by contact phenomena

Traditionally, creation of a heterostructure as an interface occurring between two different materials (dielectric–dielectric, dielectric–semiconductor, metal–dielectric, metal–semiconductor) is an essential fabrication stage of any electronic or opto-electronic device. As a result, charge transport in such devices can be greatly affected by the contact properties. There are the following contact-limited conduction mechanisms: (1) thermionic or Richardson–Schottky emission, (2) Fowler–Nordheim tunneling, (3) thermally-assisted tunneling, (4) grain boundary-limited conduction. It is important to note that the contact-limited conductivity can be analysed effectively based on an appropriate band diagram of a heterojunction.

Richardson–Schottky emission
Under the applied external electric field the energy barrier at the interface metal–insulator (see figure 1.11) is lowered by the image force.

As a result, electrons, gaining sufficient energy due to thermal activation, flow to a dielectric overcoming the barrier. This effect, called Richardson–Schottky

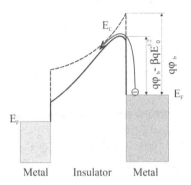

Figure 1.11. Band diagram of a metal–insulator–metal heterostructure at the Richardson–Schottky emission conditions.

emission affects current–voltage (*I–V*) characteristics that obey the following law [96]:

$$J = 2q\left(\frac{2\pi m kT}{h^2}\right)^{3/2} \mu E_0 \exp\left(-\frac{q\varphi_b}{kT}\right) \exp\left(\frac{\beta q E_0^{1/2}}{kT}\right) \qquad (1.8)$$

Here *J* is current density, *q* is the elementary charge, *m* is the electron mass, *k* is Boltzmann's constant, *T* is temperature, *h* is Plank's constant, $q\varphi_b$ is the conduction band offset (the Schottky barrier), E_0 is electric field at metal–insulator interface, μ is the carrier's mobility. Parameter β depends only on the dielectric permittivity of a material as:

$$\beta = \sqrt{\frac{q}{\pi \varepsilon \varepsilon_0}} \qquad (1.9)$$

Where ε_0 is the dielectric permittivity of free space. It follows from equation (1.8) that in the case of Richardson–Schottky emission the current–voltage characteristic is linear in $\ln(J/E_0 T^{3/2}) - E_0^{1/2}$ coordinates (the Simmons coordinates) with a slope of $q\beta/kT$. It is worth noting, the barrier height $q\varphi_b$ can be determined from the slope of a graph describing temperature dependence (in the Arrhenius coordinates) of the intercept of *I–V* characteristic in the Simmons coordinates with vertical axis. At relatively high temperatures, thermionic emission very often affects conduction mechanisms in the semiconductor heterostructures. Thus, a barrier height $q\varphi_b$ at heterojunction and dielectric permittivity of an insulator can be derived through the parameter β (see equation (1.9)).

Thermionic emission has been reported in many studies on LiNbO$_3$-based heterostructures [94, 97–99]. For instance, in a LiNbO$_3$/GaN heterostructure fabricated by the pulse-laser deposition method the potential barrier for holes with height of 0.34 eV (for applied electric field up to 4×10^6 V m^{-1}) is the limiting factor of charge transport in the framework of the Richardson–Schottky emission [98]. In LiNbO$_3$/Si heterostructures, fabricated by the RFMS method, conductivity is also limited by the thermionic emission with a barrier height of 0.2 eV for electrons, associated with a conduction band offset of 0.6 eV [99]. The dielectric constant of LiNbO$_3$ films derived experimentally from parameter β was equal to $\varepsilon = 28$ which is close to that estimated from the capacitance–voltage measurements [100, 101].

Fowler–Nordheim tunneling
In semiconductor heterostructures, when thermal emission is negligible (at relatively low temperatures), direct tunneling through a quite thin potential barrier ($\leqslant$ 50 Å) dominates. For sufficiently high applied electric fields electrons can penetrate through the triangular potential barrier for the metal–insulator–metal structure (see figure 1.12(a)).

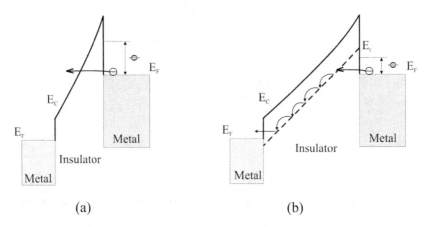

(a) (b)

Figure 1.12. Band diagram of a metal–insulator–metal system in the condition of the Fowler–Nordheim tunneling for thin insulator (a) and thick insulator (b) with the trap levels E_t below the conduction band edge E_c.

Current–voltage characteristics in this case obey the Fowler–Nordheim formula [102]:

$$J_{F-N} = AE^2 \exp\left(-\frac{B}{E}\right) \qquad (1.10)$$

Here, A and B are the constants of a material, E is electric field strength. Constant B is described by the following expression:

$$B = \frac{8\pi\sqrt{2m^*}\,\phi^{3/2}}{3qh} \qquad (1.11)$$

Here, h is Planck's constant, m^* is the electron effective mass and ϕ is the average potential barrier height. It follows from equation (1.10) that current–voltage characteristics should be a straight line in $\ln(J/E^2)$ versus $1/E$ coordinates (Fowler–Nordheim coordinates) with a slope of B. An average barrier height ϕ can be derived using equation (1.11). Although tunneling through a quite thick insulator ($d \gg 50$ Å) is improbable, this phenomenon occurs due to the presence of shallow traps in the band gap. In this case an electron is tunneling to the trap level and then can flow through the insulator via the hopping or tunneling process between traps (see figure 1.12(b)). However, the barrier height ϕ, obtained through equation (1.11) is in fact the energy of traps E_t in the band gap of an insulator (relative to the Fermi level of metal, see figure 1.12(b)).

Normally, a polycrystalline film contains electron traps in its band gap, so the tunneling process, described above, can be observed at liquid nitrogen temperatures. Indeed, in our recent work [103] we described Fowler–Nordheim tunneling at low temperatures ($T = 90$–140 K) in (001) Si–LiNbO$_3$ heterostructures, fabricated by the ion-beam sputtering method. Although the thickness of LiNbO$_3$ films was up to 1 μm, the tunneling was possible because of traps with average energy of 1.7 eV in the band gap.

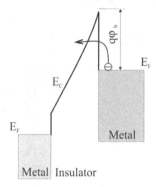

Figure 1.13. Band diagram of a metal–insulator–metal system in the condition of the thermally-assisted tunneling.

Another research group [104, 105] reported the Fowler–Nordheim emission at low temperatures in LiNbO$_3$-based heterostructures. Specifically [106], with I–V characteristics of LiNbO$_3$/AlGaN/GaN heterostructure, fabricated by the pulse-laser deposition method, clearly demonstrated the Fowler–Nordheim tunneling due to the presence of deep traps (Li vacancies) $E_t \sim 0.93$ eV in the conduction band of LiNbO$_3$.

Thermally-assisted tunneling

In the temperature range between the Fowler–Nordheim emission and Schottky emission thermally-assisted tunneling takes place. When a sufficiently strong electric field E is applied, the energy of thermally activated electrons lies between the Fermi level of a metal and the conduction band edge of an insulator. Thus the electrons are able to penetrate through the triangle potential barrier of $q\varphi_b$ (see figure 1.13).

In this case the I–V characteristic can be described by the expression [107]:

$$J = J_s\left(\exp\left(\frac{E}{E_o}\right) - 1\right) \qquad (1.12)$$

where

$$J_s = \frac{A \cdot T}{k}\sqrt{qE_{oo}\pi}\,\sqrt{\frac{q\varphi_b}{\cosh\left((E_{oo}/kT)^2\right)}}\,\exp(-q\varphi_b/E_o)$$

$$E_o = E_{oo}\coth\left(\frac{E_{oo}}{kT}\right) \qquad (1.13)$$

and

$$E_{oo} = \frac{h}{4\pi}\sqrt{\frac{N_d}{m^*\varepsilon\varepsilon_0}} \qquad (1.14)$$

Here N_d is the concentration of ionized donors in a dielectric layer, A is Richardson's constant. The thermally-assisted tunneling was observed in (001)

Si–LiNbO$_3$ heterostructures, fabricated by the ion sputtering method in an Ar atmosphere [103] and by the RFMS method in an Ar/O$_2$ gas mixture [108] with the potential barrier of 0.7 eV and 1.25 eV, respectively, at the Si/LiNbO$_3$ interface. As was mentioned above, various intrinsic and extrinsic defects are the main source of donors in LiNbO$_3$ films.

1.5.3 Electrical conductivity limited by the bulk properties

The bulk-limited conduction mechanisms are affected by bulk properties of an insulator or semiconductor and the presence of traps distributed in its band gap plays a crucial role.

This group has the following mechanisms: (1) Poole–Frenkel emission, (2) space-charge-limited currents, (3) hopping conduction, (4) grain boundary-limited conduction, (5) ionic conduction. Several very important parameters of LiNbO$_3$ films (trap density and energy distribution, the drift mobility, inter-grain barrier height, dielectric relaxation time etc) can be derived through analysing the bulk-limited conduction mechanisms.

Poole–Frenkel emission
The Poole–Frenkel effect (field-assisted thermal ionization) is the lowering of the Coulomb potential barrier when a relatively strong electric field is applied [109]. This process is the bulk analogue of the Richardson–Schottky emission at the interfacial barrier (see figure 1.14).

The current density in an insulator with shallow traps obeys the following law [109]:

$$J = J_s \exp\left(\frac{\beta_{P-F}E^{1/2}}{kT}\right) \tag{1.15}$$

Here J_s is influenced by the low field conductivity, β_{P-F} is the Poole–Frenkel coefficient, which can be defined as:

$$\beta_{P-F} = (q^3/\pi\varepsilon\varepsilon_0)^{1/2} \tag{1.16}$$

Here ε_0 is the permittivity of free space, ε is the dielectric constant of an insulator. As can be seen from equation (1.15), I–V characteristics must be linear in

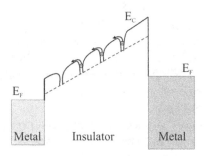

Figure 1.14. Schematic energy band diagram of the Poole–Frenkel emission in metal–insulator–metal structure.

coordinates $\ln(J) - E^{1/2}$ with a slope proportional to $\beta_{\text{P-F}}$. Trap concentration N_t can be found through the Poole–Frenkel coefficient in the following way [110]:

$$N_t = \left(\frac{q\sqrt{E_k}}{\beta_{\text{P-F}}}\right)^3 \qquad (1.17)$$

where E_k is the field corresponding to the beginning of a linear section of I–V characteristic in $\ln(J) - E^{1/2}$ coordinates. The Poole–Frenkel emission was a dominant conduction mechanism in Si–LiNbO$_3$ heterostructures, fabricated by the sol–gel technique [111], RFMS method [101] and the ion-beam sputtering method [112]. I–V analysis in the framework of Poole–Frenkel effect revealed the shallow traps in lithium niobate with concentration of $N_t = 2.4 \times 10^{17}$ cm^{-3} [101] and energy of 0.1 eV below the conduction band [112].

Space-charge-limited currents
Conduction mechanism can be the space-charge-limited (SCLC) one when ohmic contacts exist on the insulator with shallow traps. If an amount of space charge is injected in the insulator, having traps distributed in band gap, a large fraction of this charge condenses in the insulator. SCLC demonstrates specific temperature dependence because the occupancy of the traps is a strong function of temperature. I–V characteristic in this case is affected by trap distribution. In the simplest case of monoenergetic shallow traps with energy of E_t below the conduction band I–V characteristic can be expressed as [109]:

$$J = \frac{9}{8}\varepsilon\varepsilon_0\mu\theta\frac{V^2}{d^3} \qquad (1.18)$$

Here V is the applied voltage, μ is the mobility of the carriers, d is the film thickness, θ is the ratio free charge to the trapped charges. It is important to note, that when SCLC takes place, I–V characteristics should be linear in double logarithm coordinates ($\ln(I)$ versus $\ln(V)$) with a slope of 2 ('trap square law' beyond the trap filled regime) being a 'fingerprint' of this conduction mechanism. If another trap distribution exists, the I–V curve differs from the one which follows from equation (1.18) and theory of this phenomenon can be found in the excellent review [109]. Recall that regardless of fabrication methods, LiNbO$_3$ films have widely distributed traps in their band gap. Therefore, it is not surprising that SCLC is a frequently observed charge transport mechanism [97, 98, 111, 113]. Figure 1.15 demonstrates the typical I–V characteristics of LiNbO$_3$–Si heterostructures, fabricated by the RFMS method. Two specific linear sections with different slopes α (dependent on specific SCLC regimes) are clearly seen there.

The exponential trap distribution, started at 0.4 eV below the conduction band edge of LiNbO$_3$, was revealed based on I–V analysis in work [113]. Another research group reported on degenerated vacancies (V_{Li} etc), acting as traps in the band gap of LiNbO3, and filled gradually by electrons, injected from a metallic electrode [97]. I–V characteristics had several linear parts in double logarithmic coordinates, similar to those, shown in figure 1.15.

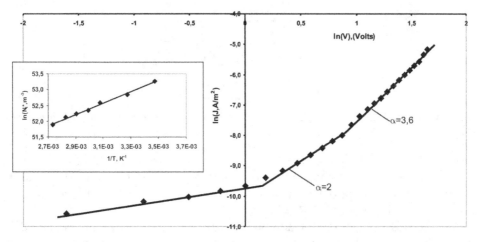

Figure 1.15. *I–V* characteristic of (001) Si–LiNbO₃ heterostructures in coordinates ln *J* – ln *V*. Temperature dependence of trap concentration N_t^* is shown in the inset [113]. Reprinted by permission from Springer Nature. Copyright 2012.

Hopping conduction
Hopping conduction is often observed in disordered materials and LiNbO₃ is not an exception. There are the following attributes of hopping conductivity:
   (a) current depends linearly on applied voltage;
   (b) relatively low activation energy of conductivity;
   (c) ac conductivity demonstrates the 'universal power law': $\sigma = A\omega^s$ (where A is a frequency independent parameter, s is a power exponent depended on a particular hopping mechanism and configuration).

Weak temperature dependence of hopping conductivity follows from the fact that it is necessary to transfer enough energy from phonons to charge carriers to overcome the energy difference between initial and final states due to applied electric field. According to Mott's model [114], if there is a spatial and energy distribution of cites near the Fermi level in the band gap of a material, an electron hops primarily between sites with the smallest energy difference and the shortest distance between them (see figure 1.16).

In the framework of Mott's model, temperature dependence of dc conductivity is described by the following expression [114]:

$$\sigma = \sigma_0(T)\exp\left(-\frac{A}{T^{1/4}}\right) \tag{1.19}$$

here $A = 2.1[\alpha^3/kN(E_F)]$. Pre-exponent factor $\sigma_0(T)$ is given as [114]:

$$\sigma_0(T) = \frac{q^2\nu_{ph}}{2(8\pi)^{1/2}}\left[\frac{N(E_F)}{\alpha kT}\right]^{1/2} \tag{1.20}$$

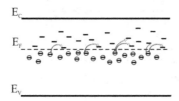

Figure 1.16. Energy diagram of a disorder semiconductor illustrating the level distribution within the band gap and the possible hops of electrons (occupied states are marked by circles).

where $N(E_F)$ is the density of electronic states, $\alpha = 1/a$ (a is the size of the localized site), ν_{ph} is the photon frequency. Two important parameters α and $N(E_F)$ can be estimated from the slope of a linear graph $\ln(\sigma T^{1/2})$ versus $T^{-1/4}$ (Mott's coordinates) and from the intercept at $T^{-1/4} = 0$.

With regards to the rise of ac conductivity with frequency, this is caused by carriers hopping back and forth over a certain number of hops demonstrating the power law: $\sigma = A\omega^s$ with a power exponent s lying between 0.5 and 1.

It was reported in [115], that despite the fact that the ac conductivity of highly oriented LiNbO$_3$ films, fabricated by the pulsed-laser deposition method, is significantly higher than the dc one, both σ_{dc} and σ_{ac} are associated with the same hopping mechanism. Due to the presence of randomly distributed defects in the band gap of LiNbO$_3$ films, electrons being the major charge carriers, facilitate the hopping charge transport with activation energy of 0.3 eV. Electronic hopping conductivity in thin LiNbO$_3$ films formed by thermal evaporation onto glass substrates was also reported in [116]. Two activation processes with activation energies of 0.067 eV and 0.57 eV were detected at temperatures below and above 423 K, respectively. In our work [101] temperature dependence of dc hopping conductivity of (001) Si–LiNbO$_3$ heterostructures fabricated by RFMS method was linear in the Mott's coordinates. We demonstrated, that dc electronic hopping conductivity with activation energy of 0.27 eV and with average hopping length $R = 126$ Å over defect centers with the concentration of $N_t = 2.5 \times 10^{17}$ cm^{-3} is the main charge transport mechanism in the studied heterostructures. In the framework of various hopping mechanisms in frequency range of 30–10^4 Hz ac conductivity of Si–LiNbO$_3$ heterostructures fabricated by RFMS method was analysed in [117]. Activated hopping transport with activation energy of 0.4 eV occurs due to the presence of defects with concentration of $N_t = 7 \times 10^{18}$ cm^{-3}.

It is important to note that if defect centers are distributed in an insulator with high density near the Fermi level, the non-activated hopping conductivity mechanism occurs at high enough applied electric fields with I–V characteristics obeying the following law [118]:

$$J = J_0 \exp(-(E_0/E)^{1/4}) \tag{1.21}$$

Here E is the applied electric field, J_0 is a field independent constant, E_0 is the characteristic field, which is expressed by the following formula:

$$E_0 = \frac{16}{D(E)a^4q} \tag{1.22}$$

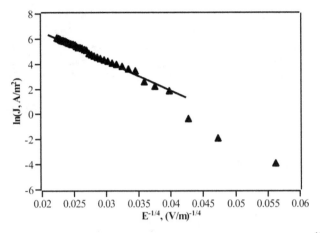

Figure 1.17. $I–V$ characteristic of LiNbO$_3$ film in coordinates $\ln(J) – E^{-1/4}$.

Here $D(E)$ is an energy density of localized states near the Fermi level, a is the localization length. As follows from equation (1.21), $I–V$ characteristics is a straight line in $\ln(J)$ versus $E^{-1/4}$ coordinates with a slope giving E_0. Indeed, in [95] we observed this conductivity mechanism in LiNbO$_3$ films with linear $I–V$ characteristics according to equation (1.21) (see figure 1.17).

Furthermore, non-activated hopping conductivity over defect centers (Li vacancies) distributed with density of $D(E) = 1.5 \times 10^{27}$ eV^{-1}·m^{-3} in the band gap of LiNbO$_3$ films fabricated by RFMS plays a dominant role in charge transport process at high electric fields ($E = 5 \times 10^5 – 3 \times 10^6$ V m^{-1}) [113, 119].

Grain boundary-limited conduction
In polycrystalline materials, the resistivity of grain boundaries can exceed the resistivity of the grains depending on various parameters like temperature, grain size and orientation etc. Therefore, in this case charge transport in the polycrystalline films is influenced by electrical properties of grain boundaries rather than a bulk material. Therefore, electrical conduction is called the grain boundary-limited conduction. Charge carriers moving from grain to grain are scattered by charged interfaces at inter-grain boundaries. On the other hand, only electrons with sufficient energy $E > \phi_b$ corresponding to their thermal velocity, can cross the barrier ϕ_b, existing at the grain boundaries (see figure 1.18).

To separate the grain boundary from a bulk contribution in total conductivity, impedance spectroscopy can be used [120]. An appropriate equivalent circuit of the studied film can help to distinguish bulk and grain boundary contributions because of different relaxation times associated with their response to an ac signal. Specifically, the simplest equivalent circuits, representing resistivity and capacitance of grain boundaries (R_{gb}, C_{gb}) and bulk of grains (R_b, C_b) in LiNbO$_3$ films is shown in figure 1.19.

The characteristic relaxation time or time constant corresponding to each parallel RC element is denoted as $\tau = RC$. A maximum in the impedance spectra occurs at

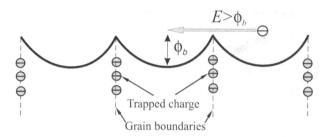

Figure 1.18. Schematic band diagram of a *n*-type polycrystalline semiconductor with grain boundaries and intergranular barriers of the height ϕ_b.

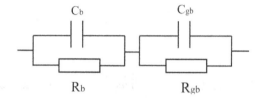

Figure 1.19. A possible equivalent circuit of LiNbO$_3$ films.

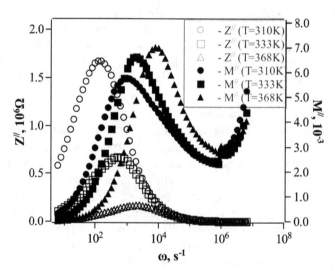

Figure 1.20. Imaginary parts of complex impedance (Z'') and dielectric modulus (M'') of as-grown (001) Si–LiNbO$_3$ heterostructures at different temperatures [121]. Reprinted by permission from Springer Nature. Copyright 2016.

characteristic frequency of $\omega_m = 1/RC$. Thus, both grain boundary and bulk parameters of a polycrystalline film can be estimated if experimental data is represented through a frequency dependence of imaginary parts of impedance ($Z''(\omega)$) and dielectric modulus ($M''(\omega)$) with a peak at frequency of ω_m. By analysing

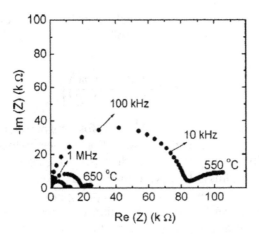

Figure 1.21. Complex impedance data at different temperatures of the LiNbO$_3$ ceramic sample [122]. Reprinted with the permission of AIP Publishing.

impedance spectra (see figure 1.20), structure and composition of thin polycrystalline LiNbO$_3$ films, grown by RFMS method, we estimated the ratio between the average depletion area width within a single grain and the size of a neutral (bulk) area $d_{gb}/d_b = 0.4$ [121]. Also, we demonstrated that the grain boundary-limited electrical conductivity of the studied films is caused by intergranular barriers of 0.4 eV. Thermal annealing of as-grown films leads to a rise in the ratio of d_{gb}/d_b up to 1, making LiNbO$_3$ films more homogeneous with an increased role of the bulk conductivity [121].

Another method to separate grain boundary and bulk contributions to a total complex impedance Z^* of LiNbO$_3$ films, deposited by a chemical evaporative technique, was given in [122]. Figure 1.21 demonstrates Nyquist plot (Im(Z^*) versus Re(Z^*) graph) with two semicircles, attributed to grain boundaries (a wide semicircle at low frequencies) and bulk of grains (a smaller semicircle at higher frequencies). This work, aiming to investigate bulk dielectric properties of polycrystalline LiNbO$_3$ films, reported thermally activated bulk conductivity with activation energy of $E_a = 1.17$ eV, which apparently results from the thermal activation of electrons from deep donors.

Since frequency response plays an important role in some applications of LiNbO$_3$ films (selective filters, electro-optical modulators, tunable microring optical resonators etc), the grain boundaries should be taken into account. We have clearly demonstrated [113, 117] that thermal annealing of LiNbO$_3$ films declines (or even suppresses) contribution of grain boundaries in total conductivity, which is responsible for a low frequency response ('slow process'), affecting LiNbO$_3$-based devices.

Ionic conduction
When various defects exist in an insulator, ions may jump from one defect site to another one under an applied electric field. The resulting process is known as the

ionic conduction. If an applied electric field satisfies the condition of $E \approx kT/qd$, current–voltage characteristics obeys the following expression for ionic conduction [102]:

$$J = C \exp\left[-\left(\frac{\phi}{kT} - \frac{Eqd}{2kT}\right)\right] \qquad (1.23)$$

Here C is a constant, ϕ is a potential barrier for ions between two sites separated by distance of d, E is the applied electric field. Some investigators suggested that due to quite high mobility of Li ions, ionic conductivity is the dominant charge transport mechanism in $LiNbO_3$ films [123, 124]. Initially, their $LiNbO_3$ films, deposited by an E-gun evaporation method, manifested the hopping electronic ac conductivity due to Li defects. When the surface of $LiNbO_3$ film was coated with a Li–Nb–O layer followed by thermal annealing, the resulting film demonstrated ionic conductivity due to diffusion of Li ions into film. Another research group, analysing dielectric relaxation and ac conductivity of congruent $LiNbO_3$ single crystals, state that thermally activated conductivity ($E_a = 0.9$ eV) is attributed to a long range motion of Li^+ ions [125]. However, in our view, there is a little evidence to support this opinion. The point is that it is hard to distinguish between the ionic and electronic conductivity in a dielectric film. If relatively high current flows through the film for a long period of time, ionic species are deposited upon electrodes, producing the material transport—one of the strongest criterions of ionic conduction. The second crucial factor associated with ionic conductivity is the polarization effects. In this case film resistivity rises with time due to the formation of space charges near the electrodes. Taking into account that these effects have not been revealed in the above mentioned works, more detailed study is needed to decide whether electronic or ionic charge transport (or both) affects conductivity of the studied films.

Summary and emphasis

The practical interest in synthesis and the properties of $LiNbO_3$ films, which has boomed in the past several decades is fully reasonable. Requirements for the structure, morphology and properties of $LiNbO_3$ films are formulated only in the 'first approximation' for existing $LiNbO_3$ films. The restrictions originate from the fact that it is impossible to form single crystal films of complex oxides even taking into account the possible synthesis of epitaxial structures within existing synthesis methods, so it is still unattainable to fabricate the films having properties close to bulk $LiNbO_3$.

In the process of RFMS, the elementary processes of stochastic growth, affected by the directional flow of ions and the wide range of their energies, force the formation of highly dispersed mosaic structures.

Thin $LiNbO_3$ films with arbitrary grain orientation, fabricated by RFMS without ion assist, manifest ferroelectric properties, but their remnant polarization is considerably lower than that for bulk lithium niobate. With that, films with single-axis texture <0001>, having formation facilitated by the ion assist, are close to bulk $LiNbO_3$ material in terms of the property.

The oriented crystallization of $LiNbO_3$ films is not clear, and few substrate-film systems have been investigated. It is worth noting, that systematic investigations suggest the possibility of growing oriented $LiNbO_3$ films for $LiNbO_3$-based heterostructures synthesized by various methods.

No systematic data has been reported regarding the structure of grain boundaries and interfaces, defects required to estimate the degree of their influence on the film properties. Nonequilibrium processes used for film fabrication (such as RFMS and PLD) imply a high degree of crystal structure imperfection.

In the last 15 years, the study of $LiNbO_3$ films has focused on their possible integration with planar technology to create new devices. There is notable progress in the study of synthesis, structure and properties of $LiNbO_3$-based heterostructures fabricated by effective methods such as RFMS and PLD.

Electrical properties of $LiNbO_3$ films are significantly influenced by deposition techniques and they depend on the synthesis condition greatly. The required characteristics and device reliability stimulates the progress in fundamental study of the growth and structure of $LiNbO_3$ films, as well as development of new synthesis methods and their optimization.

References

[1] Ballman A A 1965 Growth of piezoelectric and ferroelectric materials by the Czochralski technique *J. Am. Ceram. Soc.* **48** 112–3

[2] Abrahams S C, Levinstein H J and Reddy J M 1966 Ferroelectric lithium niobate. 5. Polycrystal x-ray diffraction study between 24° and 1200 °C *J. Phys. Chem. Solids* **27** 1019–26

[3] Abrahams S C, Reddy J M and Bernstein J L 1966 Ferroelectric lithium niobate. 3. Single crystal x-ray diffraction study at 24 °C *J. Phys. Chem. Solids* **27** 997–1012

[4] Abrahams S C and Marsh P 1986 Defect structure dependence on composition in lithium niobate *Acta Crystallogr. Sect. B Struct. Sci.* **42** 61–8

[5] Rauber A 1978 Chemistry and physics of lithium niobate *Curr. Top. Mater. Sci.* **1** 481–601

[6] Hatano H, Kitamura K and Liu Y 2007 Growth and photorefractive properties of stoichiometric $LiNbO_3$ and $LiTaO_3$ *Photorefractive Materials and Their Applications 2* (New York: Springer), pp 127–64

[7] Nassau K, Levinstein H J and Loiacono G M 1966 Ferroelectric lithium niobate. 2. Preparation of single domain crystals *J. Phys. Chem. Solids* **27** 989–96

[8] Coburn J W and Winters H F 1983 Plasma-assisted etching in microfabrication *Annu. Rev. Mater. Sci.* **13** 91–116

[9] Tamura M and Yoshikado S 2003 Etching characteristics of $LiNbO_3$ crystal by fluorine gas plasma reactive ion etching *Surf. Coatings Technol.* **169** 203–7

[10] Yang W S, Lee H-Y, Kim W K and Yoon D H 2005 Asymmetry ridge structure fabrication and reactive ion etching of $LiNbO_3$ *Opt. Mater.* **27** 1642–6

[11] Ching W Y, Gu Z-Q and Xu Y-N 1994 First-principles calculation of the electronic and optical properties of $LiNbO_3$ *Phys. Rev.* B **50** 1992–5

[12] Schmidt W G, Albrecht M, Wippermann S, Blankenburg S, Rauls E, Fuchs F, Rödl C, Furthmüller J and Hermann A 2008 $LiNbO_3$ ground- and excited-state properties from first-principles calculations *Phys. Rev.* B **77** 35106

[13] Bergman J G, Ashkin A, Ballman A A, Dziedzic J M, Levinstein H J and Smith R G 1968 Curie temperature birefringence, and phase-matching temperature variations in LiNbO$_3$ as a function of melt stoichiometry *Appl. Phys. Lett.* **12** 92

[14] Pogossian S P and Le Gall H 2003 Modeling planar leaky optical waveguides *J. Appl. Phys.* **93** 2337

[15] Fork D K, Armani-leplingard F and Kingston J J 1996 Application of electroceramic thin films to optical waveguide devices *MRS Bull.* 53–8

[16] Wohltjen H 1984 Mechanism of operation and design considerations for surface acoustic wave device vapour sensors *Sens. Actuators* **5** 307–25

[17] Runde D, Brunken S, Rüter C E and Kip D 2006 Integrated optical electric field sensor based on a Bragg grating in lithium niobate *Appl. Phys.* B **86** 91–5

[18] Lee T C, Lee J T, Robert M A, Wang S and Rabson T A 2003 Surface acoustic wave applications of lithium niobate thin films *Appl. Phys. Lett.* **82** 191–3

[19] Shih W-C, Sun X-Y, Wang T-L and Wu M-S 2009 Growth of *c*-axis oriented LiNbO$_3$ film on sapphire by pulsed laser deposition for surface acoustic wave applications *Ferroelectrics* **381** 92–9

[20] Kadota M, Ogami T, Yamamoto K, Tochishita H and Negoro Y 2010 High-frequency lamb wave device composed of MEMS structure using LiNbO$_3$ thin film and air gap *IEEE Trans. Ultrason. Ferroelectr. Freq. Control* **57** 2564–71

[21] Dogheche E, Lansiaux X, Remiens D, Sadaune V, Chauvin S and Gryba T 2003 Growth process and surface acoustic wave characteristics of LiNbO$_3$/diamond/silicon multilayered structures *Jpn. J. Appl. Phys.* **42** 572–4

[22] Tan S, Gilbert T, Hung C-Y, Schlesinger T E and Migliuolo M 1996 Sputter deposited c-oriented LiNbO$_3$ thin films on SiO$_2$ *J. Appl. Phys.* **79** 3548

[23] Gupta V, Bhattacharya P, Yuzyuk Y I, Katiyar R S, Tomar M and Sreenivas K 2004 Growth and characterization of *c*-axis oriented LiNbO$_3$ film on a transparent conducting Al:ZnO inter-layer on Si *J. Mater. Res.* **19** 2235–9

[24] Lanzhong H 2009 Epitaxial fabrication and memory effect of ferroelectric LiNbO$_3$ film/AlGaN/GaN heterostructure *Appl. Phys. Lett.* **95** 232907

[25] Scott J F 2000 Ferroelectric memories today *Ferroelectrics* **236** 247–58

[26] Wang S, Wang W, Yakopcic C, Shin E, Subramanyam G and Taha T M 2017 Experimental study of LiNbO$_3$ memristors for use in neuromorphic computing *Microelectron. Eng.* **168** 37–40

[27] Li H, Xia Y, Xu B, Guo H, Yin J and Liu Z 2010 Memristive behaviors of LiNbO$_3$ ferroelectric diodes *Appl. Phys. Lett.* **97** 12902

[28] Greenlee J D, Petersburg C F, Laws Calley W, Jaye C, Fischer D A, Alamgir F M and Alan Doolittle W 2012 *In-situ* oxygen x-ray absorption spectroscopy investigation of the resistance modulation mechanism in LiNbO$_2$ memristors *Appl. Phys. Lett.* **100** 182106

[29] Pan X *et al* 2016 Rectifying filamentary resistive switching in ion-exfoliated LiNbO$_3$ thin films *Appl. Phys. Lett.* **108** 32904

[30] Strukov D B, Snider G S, Stewart D R and Williams R S 2008 The missing memristor found *Nature* **453** 80–3

[31] Yang J J, Pickett M D, Li X, Ohlberg D A A, Stewart D R and Williams R S 2008 Memristive switching mechanism for metal/oxide/metal nanodevices *Nat. Nanotechnol.* **3** 429–33

[32] Di Ventra M, Pershin Y V and Chua L O 2009 Circuit elements with memory: memristors, memcapacitors, and meminductors *Proc. IEEE* **97** 1717–24

[33] Sheng P, Zeng F, Tang G S, Pan F, Yan W S and Hu F C 2012 Structure and ferromagnetism in vanadium-doped $LiNbO_3$ *J. Appl. Phys.* **112** 1–7

[34] Cheng, Song C, Wang X, Liu F and Zeng F P 2009 Room temperature ferromagnetism in cobalt-doped $LiNbO_3$ single crystalline films *Cryst. Growth Des.* **8** 1235–9

[35] Díaz-Moreno C, Farias R, Hurtado-Macias A, Elizalde-Galindo J and Hernandez-Paz J 2012 Multiferroic response of nanocrystalline lithium niobate *J. Appl. Phys.* **111** 5–8

[36] Ishii M, Ohta D, Uehara M and Kimishima Y 2012 Ferromagnetism of nano-$LiNbO_3$ with vacancies *Trans. Mater. Res. Soc. Jpn.* **37** 443–6

[37] Vasconcelos N S L S *et al* 2003 Epitaxial growth of $LiNbO_3$ thin films in a microwave oven *Thin Solid Films* **436** 213–9

[38] Kondo S, Miyazawa S, Fushimi S and Sugii K 1975 Liquid-phase-epitaxial growth of single-crystal $LiNbO_3$ thin film *Appl. Phys. Lett.* **26** 489

[39] Miyazawa S, Fushimi S and Kondo S 1975 Optical waveguide of $LiNbO_3$ thin film grown by liquid phase epitaxy *Appl. Phys. Lett.* **26** 8

[40] Dubs C, Ruske J-P, Kräußlich J and Tünnermann A 2009 Rib waveguides based on Zn-substituted $LiNbO_3$ films grown by liquid phase epitaxy *Opt. Mater.* **31** 1650–7

[41] Lu Y, Dekker P and Dawes J M 2009 Growth and characterization of lithium niobate planar waveguides by liquid phase epitaxy *J. Cryst. Growth* **311** 1441–5

[42] Margueron S, Bartasyte A, Plausinaitiene V, Abrutis A, Boulet P, Kubilius V and Saltyte Z 2013 Effect of deposition conditions on the stoichiometry and structural properties of $LiNbO_3$ thin films deposited by MOCVD *Proc. SPIE 8626, Oxide-based Materials and Devices IV* **8626** 862612

[43] Dislich H 1971 New routes to multicomponent oxide glasses *Angew. Chem. Int. Ed. Engl.* **10** 363–70

[44] Nashimoto K, Cima M J, McIntyre P C and Rhine W E 1995 Microstructure development of sol-gel derived epitaxial $LiNbO_3$ thin films *J. Mater. Res.* **10** 2564–72

[45] Simões A Z, Ries A, Riccardi C S, Gonzalez A H, Zaghete M A, Stojanovic B D, Cilense M and Varela J A 2004 Potassium niobate thin films prepared through polymeric precursor method *Mater. Lett.* **58** 2537–40

[46] Fakhri M A, Al-Douri Y, Hashim U, Salim E T, Prakash D and Verma K D 2015 Optical investigation of nanophotonic lithium niobate-based optical waveguide *Appl. Phys. B Lasers Opt.* **121** 107–16

[47] Schwarz H and Tourtellotte H A 1969 Vacuum deposition by high-energy laser with emphasis on barium titanate films *J. Vac. Sci. Technol.* **6** 373–8

[48] Shibata Y, Kaya K, Akashi K, Kanai M, Kawai T and Kawai S 1993 Epitaxial growth of $LiNbO_3$ films on sapphire substrates by excimer laser ablation method and their surface acoustic wave properties *Jpn. J. Appl. Phys.* **32** L745–L747

[49] Ogale S B, Nawathey-Dikshit R, Dikshit S J and Kanetkar S M 1992 Pulsed laser deposition of stoichiometric $LiNbO_3$ thin films by using O_2 and Ar gas mixtures as ambients *J. Appl. Phys.* **71** 5718

[50] Joshi V, Roy D and Mecartney M L 1993 Low temperature synthesis and properties of lithium niobate thin films *Appl. Phys. Lett.* **63** 1331

[51] Balestrino G *et al* 2001 Epitaxial $LiNbO_3$ thin films grown by pulsed laser deposition for optical waveguides *Appl. Phys. Lett.* **78** 1204–6

[52] Jelínek M, Havránek V, Remsa J, Kocourek T, Vincze A, Bruncko J, Studnička V and Rubešová K 2013 Composition, XRD and morphology study of laser prepared LiNbO$_3$ films *Appl. Phys. A Mater. Sci. Process.* **110** 883–8

[53] Wang X, Liang Y, Tian S, Man W and Jia J 2013 Oxygen pressure dependent growth of pulsed laser deposited LiNbO$_3$ films on diamond for surface acoustic wave device application *J. Cryst. Growth* **375** 73–7

[54] Wang X, Ye Z, Li G and Zhao B 2007 Influence of substrate temperature on the growth and optical waveguide properties of oriented LiNbO$_3$ thin films *J. Cryst. Growth* **306** 62–7

[55] Foster N F 1969 The deposition and piezoelectric characteristics of sputtered lithium niobate films *J. Appl. Phys.* **40** 420–1

[56] Russo D P G and Kumar C S 1973 Sputtered ferroelectric thin-film electro-optic modulator *Appl. Phys. Lett.* **23** 229–31

[57] Takada S, Ohnishi M, Hayakawa H and Mikoshiba N 1974 Optical waveguides of single-crystal LiNbO$_3$ film deposited by rf sputtering *Appl. Phys. Lett.* **24** 490–2

[58] Hewig G M, Jain K, Sequeda F O, Tom R and Wang P-W 1982 R.F. Sputtering of LiNbO$_3$ thin films *Thin Solid Films* **88** 67–74

[59] Rabson T A, Baumann R C and Rost T A 1990 Thin film lithium niobate on silicon *Ferroelectrics* **112** 265–71

[60] Kanata T, Kobayashi Y and Kubota K 1987 Epitaxial growth of LiNbO$_3$ -LiTaO$_3$ thin films on Al$_2$O$_3$ *J. Appl. Phys.* **62** 2989–93

[61] Vook R W 1984 Nucleation and growth of thin films *Opt. Eng.* **23** 343–9

[62] He J and Ye Z 2003 Highly C-axis oriented LiNbO$_3$ thin film on amorphous SiO$_2$ buffer layer and its growth mechanism *Chin. Sci. Bull.* **48** 2290

[63] Shtansky D V, Kulinich S A, Terashima K and Yoshida T 2001 Crystallography and structural evolution of LiNbO$_3$ and LiNb$_{1-x}$Ta$_x$O$_2$ films on sapphire prepared by high-rate thermal plasma spray chemical vapor deposition *J. Mater. Res.* **16** 2271–9

[64] Veignant F 1998 Epitaxial growth of LiNbO$_3$ on αAl$_2$O$_3$ (0001) *Thin Solid Films* **336** 163–7

[65] Akazawa H 2009 Target-quality dependent crystallinity of sputter-deposited LiNbO$_3$ films: Observation of impurity segregation *Thin Solid Films* **517** 5786–92

[66] Griffel G, Ruschin S and Croitoru N 1989 Linear electro-optic effect in sputtered polycrystalline LiNbO$_3$ films *Appl. Phys. Lett.* **54** 1385

[67] Park S-K, Baek M-S, Bae S-C, Kim K-W, Kwun S-Y, Kim Y-J and Kim J-H 1999 012) Preferred orientation of LiNbO$_3$ thin films by RF-magnetron sputtering *Jpn. J. Appl. Phys.* **38** 4167–71

[68] Curtis B J and Brunner H R 1975 The growth of thin films of lithium niobate by chemical vapour deposition *Mater. Res. Bull.* **10** 515–20

[69] Lee T-H, Hwang F-T, Lee C-T and Lee H-Y 2007 Investigation of LiNbO$_3$ thin films grown on Si substrate using magnetron sputter *Mater. Sci. Eng. B* **136** 92–5

[70] Shandilya S, Tomar M and Gupta V 2012 Deposition of stress free c-axis oriented LiNbO$_3$ thin film grown on (002) ZnO coated Si substrate *J. Appl. Phys.* **111** 10–6

[71] Park S K, Baek M S, Bae S C, Kwun S Y, Kim K T and Kim K W 1999 Properties of LiNbO$_3$ thin film prepared from ceramic Li-Nb-K-O target *Solid State Commun.* **111** 347–52

[72] Rost T A, Lin H, Rabson T A, Baumann R C and Callahan D L 1992 Deposition and analysis of lithium niobate and other lithium niobium oxides by rf magnetron sputtering *J. Appl. Phys.* **72** 4336–43

[73] Tsirlin M 2004 Influence of gas phase composition on the defects formation in lithium niobate *J. Mater. Sci.* **39** 3187-9

[74] Gordillo-Vázquez F J and Afonso C N 2002 Influence of Ar and O_2 atmospheres on the Li atom concentration in the plasma produced by laser ablation of $LiNbO_3$ *J. Appl. Phys.* **92** 7651

[75] Wang X, He J, Huang J, Zhao B and Ye Z 2003 Effects of oxygen pressure on the *c*-axis oriented growth of $LiNbO_3$ thin film on SiO_2/Si substrate by pulsed laser deposition *J. Mater. Sci. Lett.* **22** 225-7

[76] Bornand V and Papet P 2005 $LiNbO_3$-based ferroelectric heterostructures *J. Phys. IV* **126** 89-92

[77] Kiselev D A, Zhukov R N, Bykov A S, Voronova M I, Shcherbachev K D, Malinkovich M D and Parkhomenko Y N 2014 Effect of annealing on the structure and phase composition of thin electro-optical lithium niobate films *Inorg. Mater.* **50** 419-22

[78] Simões A Z, Zaghete M A, Stojanovic B D, Gonzalez A H, Riccardi C S, Cantoni M and Varela J A 2004 Influence of oxygen atmosphere on crystallization and properties of $LiNbO_3$ thin films *J. Eur. Ceram. Soc.* **24** 1607-13

[79] Simões A Z, Zaghete M A, Stojanovic B D, Riccardi C S, Ries A, Gonzalez A H and Varela J A 2003 $LiNbO_3$ thin films prepared through polymeric precursor method *Mater. Lett.* **57** 2333-9

[80] Bornand V, Huet I and Papet P 2002 $LiNbO_3$ thin films deposited on Si substrates: a morphological development study *Mater. Chem. Phys.* **77** 571-7

[81] Edon V, Rèmiens D and Saada S 2009 Structural, electrical and piezoelectric properties of $LiNbO_3$ thin films for surface acoustic wave resonators applications *Appl. Surf. Sci.* **256** 1455-60

[82] Simões A Z, Gonzalez A H, Zaghete M A, Stojanovic B D, Cavalheiro A A, Moeckli P, Setter N and Varela J A 2002 Influence of oxygen flow on crystallization and morphology of $LiNbO_3$ thin films *Ferroelectrics* **271** 33-8

[83] Akazawa H and Shimada M 2007 Mechanism for $LiNb_3O_8$ phase formation during thermal annealing of crystalline and amorphous $LiNbO_3$ thin films *J. Mater. Res.* **22** 1726-36

[84] Akazawa H and Shimada M 2006 Precipitation kinetics of $LiNbO_3$ and $LiNb_3O_8$ crystalline phases in thermally annealed amorphous $LiNbO_3$ thin films *Phys. Status Solidi* **203** 2823-7

[85] Volk T and Wöhlecke M 2008 *Lithium Niobate: Defects, Photorefraction and Ferroelectric Switching* (Berlin: Springer)

[86] Wong K KInstitution of Electrical Engineers and INSPEC (Information service) 2002 *Properties of Lithium Niobate* (London: INSPEC/Institution of Electrical Engineers))

[87] Wilkinson A P, Cheetham A K and Jarman R H 1993 The defect structure of congruently melting lithium niobate *J. Appl. Phys.* **74** 3080

[88] Donnerberg H, Tomlinson S M, Catlow C R A and Schirmer O F 1989 Computer-simulation studies of intrinsic defects in $LiNbO_3$ crystals *Phys. Rev. B* **40** 11909-16

[89] Schirmer O F, Imlau M, Merschjann C, Schoke B and D 2009 Electron small polarons and bipolarons in $LiNbO_3$ *J. Phys. Condens. Matter* **21** 123201

[90] Herth P, Granzow T, Schaniel D, Woike T, Imlau M and Krätzig E 2005 Evidence for light-induced hole polarons in $LiNbO_3$ *Phys. Rev. Lett.* **95** 67404

[91] Reyher H-J, Schulz R and Thiemann O 1994 Investigation of the optical-absorption bands of Nb^{4+} and Ti^{3+} in lithium niobate using magnetic circular dichroism and optically detected magnetic-resonance techniques *Phys. Rev.* B **50** 3609–19

[92] Sugak D, Zhydachevskii Y, Sugak Y, Buryy O, Ubizskii S, Solskii I, Schrader M and Becker K-D 2007 *In situ* investigation of optical absorption changes in $LiNbO_3$ during reducing/oxidizing high-temperature treatments *J. Phys. Condens. Matter* **19** 86211

[93] Schirmer O F 2006 O^- bound small polarons in oxide materials *J. Phys. Condens. Matter* **18** R667–704

[94] Lim D, Jang B, moon S, Won C and Yi J 2001 Characteristics of $LiNbO_3$ memory capacitors fabricated using a low thermal budget process *Solid State Electron.* **45** 1159–63

[95] Sumets M 2017 Charge transport in $LiNbO_3$-based heterostructures *J. Nonlinear Opt. Phys. Mater* **26** 1750011

[96] Simmons J G 1967 Poole–Frenkel effect and Schottky effect in metal-insulator-metal systems *Phys. Rev.* **155** 657–60

[97] Hao L-Z, Liu Y-J, Zhu J, Lei H-W, Liu Y-Y, Tang Z-Y, Zhang Y, Zhang W-L and Li Y-R 2011 Rectifying the current-voltage characteristics of a $LiNbO_3$ Film/GaN heterojunction *Chin. Phys. Lett.* **28** 107703

[98] Guo S M, Zhao Y G, Xiong C M and Lang P L 2006 Rectifying *I-V* characteristic of $LiNbO_3$/Nb-doped $SrTiO_3$ heterojunction *Appl. Phys. Lett.* **89** 223506

[99] Ievlev V, Sumets M, Kostyuchenko A, Ovchinnikov O, Vakhtel V and Kannykin S 2013 Band diagram of the Si-$LiNbO_3$ heterostructures grown by radio-frequency magnetron sputtering *Thin Solid Films* **542** 289–94

[100] Choi S-W, Choi Y-S, Lim D-G, moon S-I, Kim S-H, Jang B-S and Yi J 2000 Effect of RTA treatment on $LiNbO_3$ MFS memory capacitors *Korean J. Ceram.* **6** 138–42

[101] Iyevlev V, Kostyuchenko A, Sumets M and Vakhtel V 2011 Electrical and structural properties of $LiNbO_3$ films, grown by RF magnetron sputtering *J. Mater. Sci. Mater. Electron.* **22** 1258–63

[102] Maissel L I and Glang R 1970 *Handbook of Thin Film Technology* (New York: McGraw-Hill)

[103] Ievlev V, Sumets M and Kostyuchenko A 2013 Conduction mechanisms in Si-$LiNbO_3$ heterostructures grown by ion-beam sputtering method *J. Mater. Sci.* **48** 1562–70

[104] Hao L *et al* 2012 Integration and electrical properties of epitaxial $LiNbO_3$ ferroelectric film on n-type GaN semiconductor *Thin Solid Films* **520** 3035–8

[105] Akazawa H 2007 Observation of both potential barrier-type and filament-type resistance switching with sputtered $LiNbO_3$ thin films *Jpn. J. Appl. Phys.* **46** L848–50

[106] Hao L Z, Zhu J, Luo W B, Zeng H Z, Li Y R and Zhang Y 2010 Electron trap memory characteristics of $LiNbO_3$ film/AlGaN/GaN heterostructure *Appl. Phys. Lett.* **96** 32103

[107] Padovani F A and Stratton R 1966 Field and thermionic-field emission in Schottky barriers *Solid. State. Electron.* **9** 695–707

[108] Sumets M, Ievlev V, Kostyuchenko A, Vakhtel V, Kannykin S and Kobzev A 2014 Electrical properties of Si-$LiNbO_3$ heterostructures grown by radio-frequency magnetron sputtering in an Ar + O_2 environment *Thin Solid Films* **552** 32–8

[109] Simmons J G 1971 Conduction in thin dielectric films *J. Phys. D. Appl. Phys.* **4** 202

[110] Hill R M 1971 Poole–Frenkel conduction in amorphous solids *Philos. Mag.* **23** 59–86

[111] Joshi V, Roy D and Mecartney M L 1995 Nonlinear conduction in textured and non textured lithium niobate thin films *Integr. Ferroelectr.* **6** 321–7

[112] Iyevlev V, Kostyuchenko A and Sumets M 2011 Fabrication, substructure and properties of LiNbO$_3$ films *Proc. SPIE* **7747** 77471J-1

[113] Iyevlev V, Sumets M and Kostyuchenko A 2012 Current–voltage characteristics and impedance spectroscopy of LiNbO$_3$ films grown by RF magnetron sputtering *J. Mater. Sci. Mater. Electron* **23** 913–20

[114] Mott N F 1969 Conduction in non-crystalline materials *Philos. Mag.* **19** 835–52

[115] Shandilya S, Tomar M, Sreenivas K and Gupta V 2009 Purely hopping conduction in c-axis oriented LiNbO$_3$ thin films *J. Appl. Phys.* **105** 94105

[116] Easwaran N, Balasubramanian C, Narayandass S A K and Mangalaraj D 1992 Dielectric and AC conduction properties of thermally evaporated lithium niobate thin films *Phys. Status Solidi* **129** 443–51

[117] Ievlev V, Sumets M, Kostyuchenko A and Bezryadin N 2013 Dielectric losses and ac conductivity of Si-LiNbO$_3$ heterostructures grown by the RF magnetron sputtering method *J. Mater. Sci. Mater. Electron.* **24** 1651–7

[118] Lösche A, Mott N F and Davis E A 1972 Electronic Processes in Non-crystalline Materials (Oxford: Clarendon Press) *Krist. Tech.* **7** K55–6

[119] Sumets M, Ievlev V, Kostyuchenko A, Kuz'mina V and Kuzmina V 2014 Influence sputtering conditions on electrical characteristics of Si-LiNbO$_3$ heterostructures formed by radio-frequency magnetron sputtering *Mol. Cryst. Liq. Cryst.* **603** 202–15

[120] Macdonald J R 1992 Impedance spectroscopy *Ann. Biomed. Eng.* **20** 289–305

[121] Sumets M, Kostyuchenko A, Ievlev V and Dybov V 2016 Electrical properties of phase formation in LiNbO$_3$ films grown by radio-frequency magnetron sputtering method *J. Mater. Sci. Mater. Electron.* **27** 7979–86

[122] Lanfredi S and Rodrigues A C M 1999 Impedance spectroscopy study of the electrical conductivity and dielectric constant of polycrystalline LiNbO$_3$ *J. Appl. Phys.* **86** 2215–9

[123] Graça M P F, Prezas P R, Costa M M and Valente M A 2012 Structural and dielectric characterization of LiNbO$_3$ nano-size powders obtained by Pechini method *J. Sol-Gel Sci. Technol.* **64** 78–85

[124] Perentzis G, Horopanitis E, Pavlidou E and Papadimitriou L 2004 Thermally activated ionic conduction in LiNbO$_3$ electrolyte thin films *Mater. Sci. Eng.* B **108** 174–8

[125] Chen R H, Chen L-F and Chia C-T 2007 Impedance spectroscopic studies on congruent LiNbO$_3$ single crystal *J. Phys. Condens. Matter* **19** 86225

Chapter 2

Synthesis, structure and surface morphology of LiNbO$_3$ films

2.1 Technological regimes of the synthesis of thin LiNbO$_3$ films by radio-frequency magnetron sputtering and ion-beam sputtering methods

Radio-frequency magnetron sputtering (RFMS) and ion beam sputtering (IBS) was performed in the sputtering system schematically shown in figure 2.1.

Single crystal lithium niobate wafers with a diameter of 65 mm were used as targets[1].

To study the effect of synthesis conditions on the structure and properties of deposited films, various sputtering parameters, given in table 2.1, were used.

Figure 2.2 shows the diagram of the relatives position of the target and a substrate in the 'offset from the target erosion zone' regime.

The optimal magnetron power was chosen according to recommendations given in chapter 1.

The sample preparation for the Si wafers and Si–SiO$_2$ heterostructures was as follows. It is known, that the use of mechanical polishing methods for sample surfaces leads to significant disruption of the crystal structure up to 50 μm in depth. To remove this disrupted layer wet-etching was used. The most common etching solution H$_2$SO$_4$:H$_2$O$_2$:H$_2$O=5:1:1 allows a high quality mirror surface to be obtained on the substrate. Nevertheless we used a plasma etching approach, to achieve the same effect in a single cycle. The substrates were etched in the same sputtering system in an Ar plasma for two minutes. This method results in the removal of the disrupted silicon layer and surface contaminations if Si–SiO$_2$ substrates are used. Furthermore, the thin SiO$_2$ film, existing on the Si substrate influences subsequent technological

---

[1] Single crystals were grown in the I V Tananaev Institute of Chemistry and Technology of Rare Elements and Mineral Raw Materials of the Russian Academy of Sciences Kola Science Center.

doi:10.1088/978-0-7503-1729-0ch2

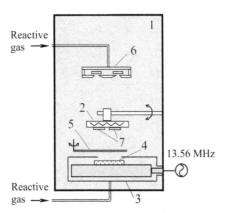

Figure 2.1. Schematic diagram of a multifunctional sputtering system: 1—reactive chamber, 2—substrate holder, 3—magnetron, 4—water cooled target, 5—shutter, 6—ion source, 7—substrate.

processes and affects electrical parameters of synthesized $LiNbO_3$-based heterostructures. This issue was also resolved by plasma etching.

The thickness control for the deposited films was performed using an AFM-profile for a control sample, fabricated under the same conditions. Thermal annealing (TA) of the studied heterostructures was conducted in a coaxial furnace in air at a temperature of 650 °C for one hour.

The study of the structure, elemental compositions and surface morphology was conducted using the methods of transmission electron microscopy (TEM, EMV-100BR), atomic force microscopy (AFM, Solver P47), Rutherford back-scattering spectrometry (RBS, α-particles with energy of 2.3 MeV), Raman spectroscopy and the x-ray diffraction method (XRD, ARL X'TRA Thermo Techno with a Cu Kα source operated at 40 kV and 35 mA). The study of the cross-sections of the specimens' heterostructure was performed with the use of a Philips EM-430 ST microscope.

Raman spectra were obtained[2] using a multi-channel spectrometer in the reflection mode (Ar laser with $\lambda = 514.5$ nm). The accuracy of measurements was ± 2 cm^{-1}.

It was necessary to separate a film from the Si substrate for transmission electron microscopy. Wet etching in the mixture of hydrofluoric and nitric acids (1:5) was used for this purpose. For the Ag substrates we used a 50% solution of nitric acid.

2.2 Composition, structure and surface morphology of LiNbO₃ films

At the initial stage of our study we estimated the potential possibility of the deposition of single phase $LiNbO_3$ films onto substrates, listed in table 2.1. Epitaxial Ag films with thickness of up to 1 μm were deposited by thermal evaporation and condensation in a vacuum (8×10^{-4} Pa) onto the heated (450 °C) fluorphlogopite plates.

---

[2] In the Institute of Spectroscopy of the Russian Academy of Sciences.

Lithium Niobate-Based Heterostructures

Table 2.1. Technological and sputtering parameters, used at the initial stage of our study.

Type of a substrate	Magnetron (supply) power	Working pressure, Pa	Reactive gas composition	Substrate temperature (°C)	Substrate position
• (001)Si (n-type conductivity with ρ = 4.5 Ω cm and p-type conductivity with ρ = 20 Ω cm) • (111)Si • (001)Si-SiO$_2$ • Fluorphlogopite • Fluorphlogopite-epitaxial Ag film	• 100 W (2k W)	• 1.5 × 10^{-1} ('low' pressure) • 5 × 10^{-1} ('high' pressure)	• Ar • Ar + O$_2$ (at the ratio of: Ar/O$_2$=60/40 Ar/O$_2$=80/20	• Unheated • 550	• Coaxially with a target ('over the target erosion zone') at different distances; • Shifted along the horizontal plane ('offset from the target erosion zone')

2-3

2.2.1 Films, deposited by the RFMS method

The films with thickness up to 4 µm were deposited by RFMS in an Ar environment at working pressure of 0.1 Pa and magnetron power of 100 W. When the substrate–target distance was equal to 4–5 cm, the growth rate was around 10 and 17 nm min^{-1}, respectively.

One of the most important properties of RFMS is the effect of spatial plasma inhomogeneity in the synthesis process. This phenomenon occurs when the composition of the films and the texture depends on the relative substrate–target position.

To estimate the degree of this effect we studied thin (0.1 µm) films, deposited at the substrate temperature of $T = 550$ °C onto (111)Si and (001)Si wafers and onto the surface of SiO_2 in the following substrate positions relative to the target erosion zone: positions 1 and 2—over the target erosion zone at a distance of 4 cm and 10 cm, respectively; positions 3 and 4—offset from the target erosion zone in the same plane.

Based on TEM patterns at the working power of 100 W the amorphous films were formed on the unheated substrates ($T = 20$–300 °C). Under electron beam exposure (in the electron microscope) these films were crystallized with the formation of nanocrystalline $LiNbO_3$ films, composed of grains sized up to 50 nm and having the preferable (110) orientation parallel to a substrate surface (see figure 2.3).

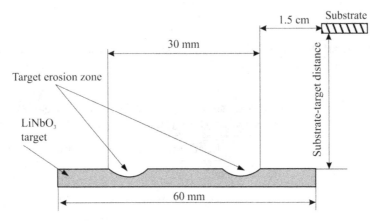

Figure 2.2. The schematic diagram of a relative target–substrate position in the 'offset from the target erosion zone' regime.

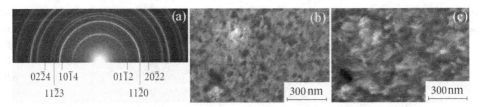

Figure 2.3. TEM patterns of $LiNbO_3$ films: (a) TEM diffraction pattern, (b) the bright-field TEM image and (c) the dark-field TEM image.

The composition of the thick films (up to 4 μm), deposited at a temperature of 550 °C onto various substrates, located over the target erosion zone, is given in figure 2.4 and table 2.2.

As follows from table 2.2, the elemental composition in the bulk of films corresponds to the stoichiometric lithium niobate (LiNbO$_3$). Also, for the films synthesized on the substrates in position A, carbon is present in the near-surface area, which is typical for the films, deposited in sputtering systems with steam-oil pumping.

Figure 2.5 demonstrates electron diffraction patterns and micrographs of the films with thickness of 0.1 μm, deposited at a temperature of $T = 550$ °C: (a)–(d) on the Si wafers (in position 1); (e), (f) on the Si–SiO$_2$ heterostructures (in position 4). Analysis of electron diffraction patterns (a) and (c) in figure 2.5 indicates that single

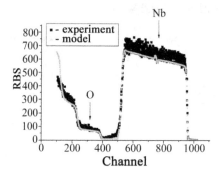

Figure 2.4. RBS spectra for the film with the thickness of 2 μm, deposited on the Si-SiO$_2$ substrate [1]. Reprinted by permission from Springer Nature. Copyright 2011.

Table 2.2. Elemental composition of films, deposited by RFMS on the (001)Si wafers (A), (111)Si-SiO$_2$ heterostructures (B) and fluorphlogopite-epitaxial (111)Ag film heterostructures (C).

Substrate	Depth (nm)	Relative concentration of elements, (at. %)						
		Li	Nb	O	Si	C	Ag	K
A	566	20.0	18.0	52.00	0.0	10.0	0.0	0.0
	1216	20.0	20.0	60.0	0.0	0.0	0.0	0.0
	1888	20.0	20.0	60.0	0.0	0.0	0.0	0.0
	1954	0.0	0.0	0.0	100.0	0.0	0.0	0.0
	42 139	0.0	0.0	0.0	100.0	0.0	0.0	0.0
B	216	20.0	20.0	60.0	0.0	0.0	0.0	0.0
	650	20.0	20.0	60.0	0.0	0.0	0.0	0.0
	1257	0.0	0.0	66.6	33.3	0.0	0.0	0.0
	19 439	0.0	0.0	0.0	100.0	0.0	0.0	0.0
C	1712	20.0	20.0	60.0	0.0	0.0	0.0	0.0
	1747	0.0	50.0	0.0	0.0	0.0	50.0	0.0
	2153	0.0	0.0	0.0	0.0	0.0	100.0	0.0
	13 646	0.0	0.0	59.0	34.7	0.0	0.0	6.2

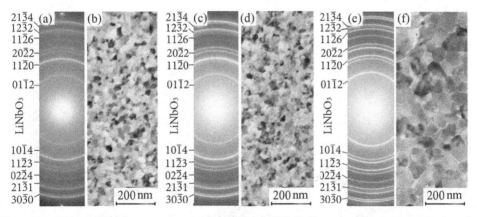

Figure 2.5. Electron diffraction patterns and micrographs of LiNbO₃ films substructure. (a) and (b) show the films deposited on the (111)Si substrates (in position 1); (c) and (d) show the films, deposited on the (001)Si substrates (in position 1); (e) and (f) show the films, deposited on (111)Si-SiO₂ heterostructure (in position 4).

phase LiNbO₃ films without any evidence of epitaxial growth are formed under these sputtering conditions. The relatively intensive reflex $11\bar{2}0$ indicates that the single axis <0001>-textured LiNbO₃ films are formed on the (001)Si and (111)Si substrates.

As follows from figure 2.5, the substructure of LiNbO₃ films, deposited onto Si substrates of both (111) and (001) orientations, is identical. It can be caused by the presence of a thin natural SiO₂ layer on the Si surface, passivating the orienting effect from a substrate. Note that the sizes of grains and sub-grains lie in the range of 10–50 nm, and the largest grains correspond to the films, synthesized on the (111)Si-SiO₂ heterostructures offset from the target erosion zone. These single-phase LiNbO₃ films with an arbitrary grain orientation indicate that the thermal activation at a given temperature ($T = 550\ °C$) is not sufficient for reactions at the SiO₂/LiNbO₃ interface.

Figure 2.6 shows XRD patterns of the films with thickness of around 1 μm on (001)Si and with thickness of about 2 μm on the (001)Si-SiO₂ heterostructure. All peaks in these patterns correspond to LiNbO₃ and the pronounced <0001> texture evolves with increasing thickness.

The fact that only peaks corresponding to LiNbO₃, with preferable orientations (001) and (104) (according to TEM data), agree with other results [2, 3].

XRD patterns of the films deposited on the (111)Ag substrate demonstrate peaks corresponding to LiNbO₃ with the (018) preferable orientation (see figure 2.7).

Figure 2.8 demonstrates the bright-field cross-sections of the (111)Si-LiNbO₃ heterostructure (a) (thickness of the LiNbO₃ films is 2.0 μm) along with the high resolution structure (b). The LiNbO₃ film is clearly seen in this figure along with the (111)Si substrate for which the interplanar distance is $d_{111} = 3.21$ Å. Regarding the atomic planes, clearly seen in the film, according to the Crystallographic database (see the Appendix) they correspond to the following planes: $(01\bar{1}2)$LiNbO₃ ($d = 3.74$ Å) and $(02\bar{2}4)$LiNbO₃ ($d = 1.87$ Å). Furthermore, as seen

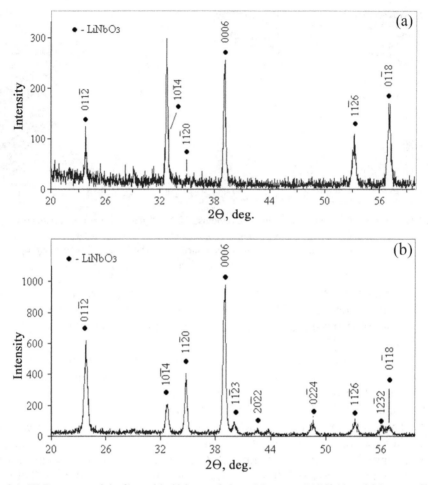

Figure 2.6. XRD patterns of the films with thickness of about 1.0 μm on (001)Si (a) and 2.0 μm on Si–SiO$_2$ (b) [1]. Reprinted by permission from Springer Nature. Copyright 2011.

in figure 2.8, a thin (5 nm) amorphous SiO$_2$ layer exists at the interface, which passivates the orienting effect of (111)Si and (001)Si substrates on the growth of LiNbO$_3$ films.

Figure 2.9 presents the bright-field cross-section patterns of (111)Ag-LiNbO$_3$ heterostructure (a) and (b) and diffraction patterns of the Ag substrate (c) and the LiNbO$_3$ film (d) for parallel zones <110> and <11$\bar{2}$0>, respectively. In this sputtering regime the two-axis (epitaxial) texture is formed with the orientation ratio of (0001), [11$\bar{2}$0] LiNbO$_3$ ∥ (111), <110> Ag, ensuring the mosaic substructure of fabricated LiNbO$_3$ films. Thus, it is advisable to use the epitaxial Ag films as substrates for deposition of LiNbO$_3$ films (and specifically for the synthesis of the films separated from a substrate).

In the following stage we studied surface morphology of as-grown films. Figure 2.10 shows AFM images obtained in the topography mode and height distribution curves for the LiNbO$_3$ films with thickness of 1 μm, deposited on (001)Si substrates

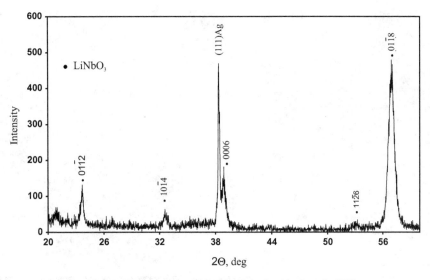

Figure 2.7. XRD pattern of LiNbO₃ films deposited on Ag(111).

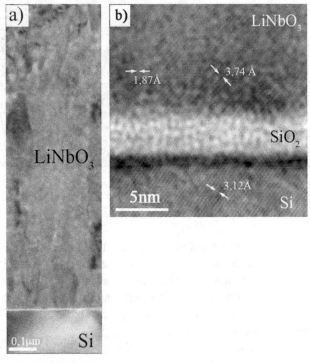

Figure 2.8. Bright-field photomicrographs of the cross-section pattern of (111)Si-LiNbO₃ heterostructures (a) (thickness of the LiNbO₃ films is 2.0 μm) along with its high resolution structure (b) [1]. Reprinted by permission from Springer Nature. Copyright 2011.

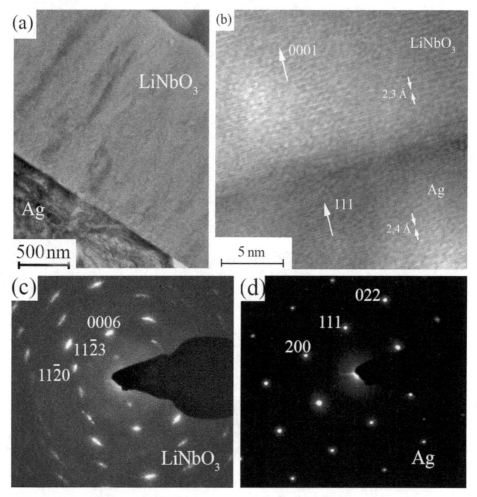

Figure 2.9. Bright-field photomicrographs of the cross-section patterns of (111)Ag-LiNbO₃ heterostructure ((a) and (b)), and the electron diffraction patterns ((c) and (d)) of the LiNbO$_3$ films (c) for <110> and <11$\bar{2}$0 > parallel zones and of the (111)Ag film (d) [4]. Reprinted by permission from Springer Nature. Copyright 2015.

at temperature of $T = 550$ °C. Analysis of these patterns has revealed that a lateral surface relief corresponds to the grain substructure of the studied films. As follows from the comparison with figure 2.8(b), this structure does not change when the thickness increases from 0.1 to 1.0 μm. It indicates that the films synthesized in this sputtering regime are not single-oriented throughout the thickness. The average size of inhomogeneities is 25 nm.

As regards films, deposited onto the Si–SiO₂ substrates, the analysis of AFM patterns (see figure 2.11) indicates that the lateral size of ingomogeneities is about 70 nm, which coincides with TEM study (see figure 2.5). Average roughness of the films is 16 nm.

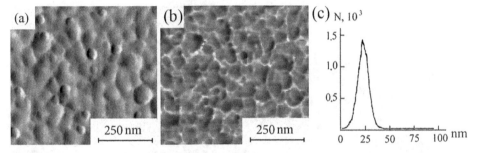

Figure 2.10. AFM images, obtained during the phase contrast mode (a), registration of the change of resonance frequency oscillations of a cantilever (b) and the height distribution curves (c), showing surface roughness of LiNbO₃ films, deposited on the (001)Si substrates by RFMS method [5]. Reprinted by permission from Springer Nature. Copyright 2017.

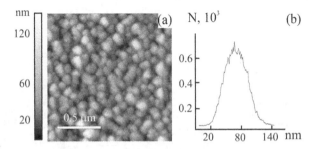

Figure 2.11. AFM images showing the surface morphology (a) and the height distribution curve (b) for the LiNbO₃ film with a thickness of 1 μm, deposited onto the Si–SiO₂ substrate offset from the target erosion zone.

The Raman spectra of the films with a thickness of 1.5 μm, deposited on the (001) Si (a) and Si–SiO₂ (b) substrates at temperature of $T = 550$ °C.

All spectra correspond to the Raman spectra of LiNbO₃ [5]. Spectral lines with frequencies of 153, 233, 264, 314, 361, 429, 589, 630, 778, 867 cm⁻¹ correspond to polycrystalline lithium niobate, fabricated by high temperature synthesis. However, almost all observed lines have lower frequencies (by a few cm⁻¹) compared to the spectra, attributed to poly- and single-crystal lithium niobite, which can be associated with deviation from a stochiometry and with the film structure.

The lines with frequencies of 518, 956, 1125 cm⁻¹ correspond to the Raman spectra of silicon, whereas a line with frequency of 956 cm⁻¹ can be attributed to the valence bridge oscillations of oxygen atoms in the 'partially destroyed' NbO₆ octahedra, formed during RFMS. Evidently, a line with frequency of 867 cm⁻¹ is doubled, which can be caused by the difference in position of Nb atoms in different octahedra.

2.2.2 Films, deposited by the IBS method

Thin films with a thickness of up to 1 μm were deposited on the (001)Si wafers by the IBS technique in Ar atmosphere under the gas pressure of 1×10^{-1} Pa and working

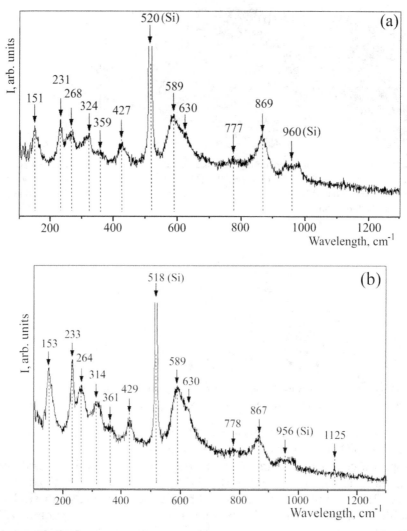

Figure 2.12. The Raman spectra of the films with a thickness of 1.5μm, grown on (001)Si (a) and Si–SiO$_2$ (b) at temperature of $T = 550$ °C.

power of 2 kW. In this regime the film growth rate is 3 nm min^{-1} at a distance of 4.0–5.0 cm between a substrate and a target.

The elemental composition of synthesized films, obtained through the RBS spectra, is given in table 2.3.

As follows from the results, the O/Nb ratio in the studied films is lower than for stochiometric lithium niobate, which indicates a relatively high concentration of oxygen vacancies in the films fabricated by the IBS method. As was demonstrated in our work, these vacancies affect electrical conductivity of LiNbO$_3$-based hetero-structures [6]. Furthermore, it follows from table 2.3 that an extended intermediate layer with variable composition is formed at the substrate/film interface. The presence of W atoms in the film is attributed to the use of a tungsten spiral to

Table 2.3. Elemental composition of the films, deposited by the IBS method on the (001)Si wafers.

Layer depth (nm)	Relative concentration of elements, (%)					
	Li	Nb	O	Si	C	W
51	21.5	17.3	43	0	18	0.3
910	26.2	21.1	52.4	0	0	0.3
1232	25.3	21.7	52.7	0	0	0.3
1555	26.5	20.3	52.9	0	0	0.3
1662	22.6	20.2	53	3.9	0	0.3
1748	19.9	15	49.9	15	0	0.3
1853	12.2	5.9	30.6	50.9	0	0.3
1976	6	2.2	11.1	80.4	0	0.3
16 843	0	0	0	100	0	0

generate thermo-electronic emission to reduce positive charge on a target during the IBS process.

A TEM study demonstrates that, similar to the films deposited by RFMS method, the films synthesized by IBS are the amorphous ones. The stable formation of nano-crystalline films is feasible only at temperatures not lower than 550 °C. Figure 2.13 demonstrates TEM patterns of the films with a thickness of 0.1 μm, deposited onto the (001)Si substrates at a temperature of 550 °C.

It follows from an electron diffraction pattern (see figure 2.13(b)), that single-phase $LiNbO_3$ films with an arbitrary grain orientation are formed in the studied sputtering conditions. As seen from the micrograph (figure 2.13(a)), the average grain size does not exceed 50 nm. XRD patterns, shown in figure 2.14, also demonstrate the single-phase nature of the studied $LiNbO_3$ films. All peaks on this diffraction pattern belong to polycrystalline lithium niobate with an average grain size of 50 nm according to calculations based on the Debye–Scherrer method.

Figure 2.15 shows AFM images of the surface and bar graphs, illustrating the heights distribution for as-grown $LiNbO_3$ films on (001)Si. The average surface roughness within the scanned area of 2.0×2.0 μm^2 is 40 nm and the average lateral size of crystalline blocks is around 200 nm. However, taking into account that the maximum height difference is considerably lower than the integral thickness of the films (see figure 2.16), we can conclude that $LiNbO_3$ films, fabricated by IBS are continuous ones with a high porosity.

Figure 2.17 shows the TEM images of the films with a thickness of 0.1 μm, deposited on the heated ($T = 550$ °C) fluorphlogopite-epitaxial (111)Ag film heterostructure.

It follows from the diffraction pattern that the Li_3NbO_4 phase along with the $LiNbO_3$ phase is formed in the films, deposited by IBS on the (111)Ag epitaxial film. The relatively high intensity of the $11\bar{2}0$ reflex indicates the formation of the <0001>-texture.

Thus, based on the results of preliminary study we can conclude that both RFMS and IBS are the effective deposition methods for $LiNbO_3$ films which preserve the

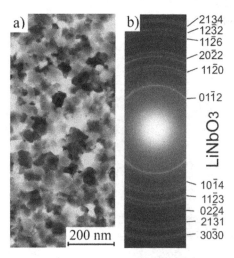

Figure 2.13. Micrographs (a) and electron diffraction patterns (b) of LiNbO₃ films with thickness of about 0.1 μm, grown on the heated (550 °C) (001)Si substrate by the IBS method [6]. Reprinted by permission from Springer Nature. Copyright 2013.

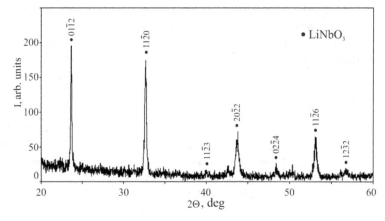

Figure 2.14. XRD pattern of the films with a thickness of 1 μm, deposited by the IBS method on the (001)Si substrates at $T = 550$ °C.

elemental composition. Thin polycrystalline LiNbO₃ films with arbitrary oriented grains 50 nm sized are formed on the heated ($T = 550$ °C) (001)Si substrates by the IBS method. At a distance of 4–5 cm between a target and a substrate the growth rate is 3 nm min⁻¹.

As regards the RFMS method, at the same substrate–target distance and the working power of 100 W, the growth rate is 10 nm min⁻¹. On the (001)Si and (111)Si substrates, the single-phase <0001>-textured LiNbO₃ films with a grain size of 10 nm are formed over the target erosion zone. The oriented effect of a substrate is

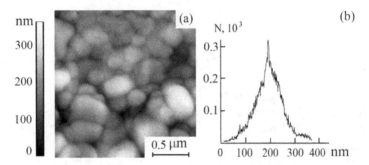

Figure 2.15. AFM images obtained during topography modes (a) and height distribution curves, showing surface roughness (b) of LiNbO$_3$ films with a thickness of 1.0 μm, formed on (001)Si substrate [6]. Reprinted by permission from Springer Nature. Copyright 2013.

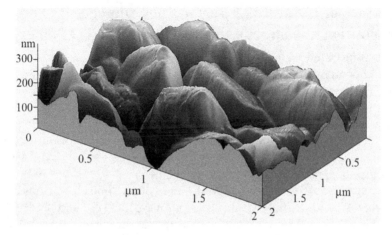

Figure 2.16. 3D AFM image of the film surface, fabricated by the IBS method.

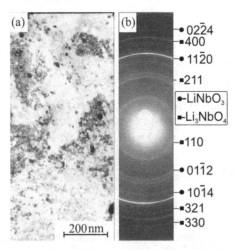

Figure 2.17. Micrographs (a) and electron diffraction patterns (b) of LiNbO$_3$ films with the thickness of about 0.1 μm, grown on the heated (550 °C) fluorphlogopite-epitaxial (111)Ag film substrate by the IBS method.

not observed due to formation of a thin (5 nm) amorphous SiO_2 layer at the substrate–film interface.

The single-phase $LiNbO_3$ films with randomly oriented grains (50 nm in size) are formed on the substrates over the target erosion zone.

At the same time, epitaxial $LiNbO_3$ films are fabricated on the (111)Ag substrates by the RFMS method. The two-axis (epitaxial) texture is formed with the orientation ratio of (0001), [11$\bar{2}$0] $LiNbO_3$ ∥ (111), <110> Ag, ensuring the mosaic substructure of fabricated $LiNbO_3$ films. However, the films, deposited by the IBS technique do not have these epitaxial properties and have the 'parasitic' Li_3NbO_4 phase.

These differences in properties indicate the necessity to study the technological regimes, effect on structure, composition and surface morphology of the synthesized $LiNbO_3$ films.

2.3 Influence of the synthesis regimes and subsequent annealing on composition and structural properties of $LiNbO_3$ films

As was demonstrated in chapter 1, sputtering conditions such as reactive gas pressure and composition, the temperature of a substrate, its position relative to a target etc, influence the composition and structure of deposited films greatly. The complication is that different combinations of technological parameters can influence properties of synthesized films in different ways, making these parameters unique. In this section we estimate the influence of the most important technological factors on structure, composition and surface morphology of synthesized films aiming to propose the optimal sputtering regimes.

We have synthesized films by the RFMS and IBS methods in various regimes on the following heated (550 °C) substrates: (001)Si, (111)Si and fluorphlogopite-epitaxial (111)Ag film heterostructure. Due to the strong dependence of plasma properties on the relative position between a substrate and a target we have used the following two basic substrate positions: 'over the target erosion zone' and 'offset the target erosion zone' with a variation of the vertical distance. This change in plasma conditions will demonstrate the 'ion assist effect' on film properties. Deposition regimes in the framework of this study are given in table 2.4.

2.3.1 Plasma effect (the ion assist effect)

First we will study the influence of an Ar plasma with a constant composition, we will then discuss the influence of oxygen on a plasma and consequently on films properties.

Figure 2.18 shows typical XRD patterns of the films, fabricated by RFMS methods in regimes 1, 2 and 3 (see table 2.4) [4].

As follows from figure 2.18, single-phase c-oriented $LiNbO_3$ films are formed under regimes 1 and 2. However, at the more intense plasma influence, when the substrates are over the target erosion zone (regime 1), this effect is not so pronounced compared to the films, deposited at the moderate plasma effect. These results agree with the data of other investigators [7]. At the same time, the films synthesized on the substrates, offset from the target erosion zone, have undergone the minimal plasma effect and do not demonstrate directional growth (see figure 2.18, regime 3). Similarly,

Table 2.4. Deposition regimes in the study of plasma effects on structure and composition of LiNbO$_3$ films.

Regime #	Deposition method	Magnetron (supply) power	Reactive gas pressure (Pa)	Substrate-target distance (cm)	Substrate position
1				5	Over the target erosion zone
2	RFMS	100 W	1.5×10^{-1}	5	Offset from the target erosion zone
3				10	Offset from the target erosion zone
4			5×10^{-1}	5	Offset from the target erosion zone
5	IBS	2 kW	1×10^{-1}	5	

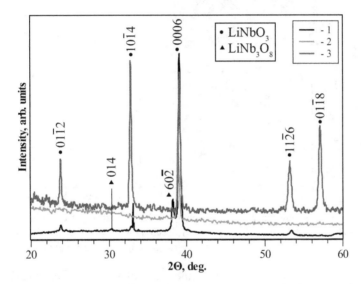

Figure 2.18. XRD patterns of the films, deposited by RFMS on the (001)Si substrates under three regimes (see table 2.4). 1: regime 1, 2: regime 2 and 3: regime 3.

the films, fabricated by IBS method (regime 5) also manifest an arbitrary grain orientation (see figure 2.14), which indicates a weak plasma effect. Thus, removal of the substrates from the target erosion zone or the use of the IBS method leads to elimination of the plasma effect on the properties of films, decreasing their growth rate.

Figure 2.19 demonstrates AFM images of morphology of the films, deposited by RFMS under regimes 2, 3 and 5 [4].

Analysis of the surface morphology demonstrates that average surface roughness increases from 10 nm (regime 1) to 30 nm (regime 5) with the decrease of the plasma effect, which is consistent with our recent paper [4].

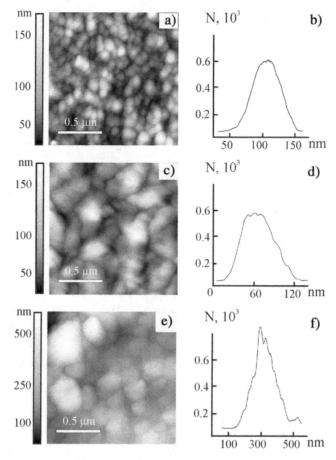

Figure 2.19. AFM images obtained during topography modes (a) and height distribution curves showing surface roughness (b) of LiNbO$_3$ films, formed on (001)Si substrates under conditions 2 (a) and (b), 3 (c) and (d) and 5 (e) and (f) (see table 2.4) [4]. Reprinted by permission from Springer Nature. Copyright 2015.

TEM patterns of the films with a thickness of 0.1 μm, fabricated by RFMS method under regimes 1(a) and (b), 2(c) and (d) and 3(e) and (f).

All reflexes in electron diffraction patterns are attributed to LiNbO$_3$. The relatively high intensity of the (01$\bar{1}$2) reflex in figure 2.20(a) and (c) indicates that a single-axis <0001> texture is formed under regimes 1 and 2. By contrast, electron diffraction patterns of the films, deposited under minimal plasma effect (regime 3), manifest an arbitrary grain orientation. Micrographs of a film's substructure (see figure 2.20(b), (d), (f)) demonstrate the crystallite size increases from 20 nm (regime 1) to 50 nm (regime 3). Thus, an increase in plasma effect leads to a decrease in the average size of crystallites and the formation of a single-axis <0001> textured LiNbO$_3$ films, as was shown in our previous works [4, 7, 8].

RBS data, reported in [4] indicates that stoichiometric LiNbO$_3$ films are formed under regime 3 (see table 2.4) with the formation of a thin intermediate layer at the Si/LiNbO$_3$ interface with a thickness of 20 nm which is apparently caused by the

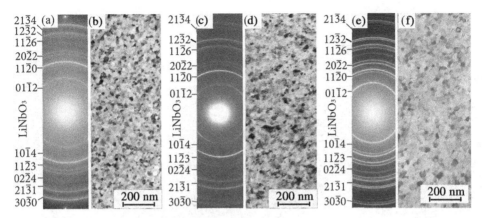

Figure 2.20. Electron diffraction patterns (a), (c), (e) and micrographs (b), (d), (f) of LiNbO$_3$ films with a thickness of 0.1 μm, grown on the (001)Si substrate under regime 1 (a), (b), regime 2 (c), (d) and regime 3 (e), (f).

diffusion of elements at that temperature. Thus, the plasma effect does not influence elemental composition of LiNbO$_3$ films, deposited by the RFMS method under the studied sputtering regimes. From this point of view, the use of this method is more favorable compared to the IBS method which does not sustain the formation of stoichiometric LiNbO$_3$ films (see table 2.3).

2.3.2 Effect of reactive gas pressure

To study this effect we have investigated films, fabricated by RFMS technique on the substrates, situated according to regime 2 in table 2.4 at different reactive gas (Ar) pressure. Figure 2.21 demonstrates the XRD pattern of the films with a thickness of 0.5 μm, fabricated by RFMS at higher argon pressure in a reactive chamber (regime 4 in table 2.4).

As follows from figure 2.21, there is a five-fold increase in reactive gas pressure compared to regime 2, which results in the formation of films containing two phases: LiNbO$_3$ and LiNb$_3$O$_8$. The lithium deficit phase LiNb$_3$O$_8$ (lithium triniobate) is a non-ferroelectric phase (often called 'parasitic'). As was stressed in chapter 1, this phase is formed when the lithium concentration in a plasma or in a film is lowered. At high working pressure in a reactive chamber Li atoms are more volatile compared to niobium [9, 10]. Furthermore, at higher working pressure the mean free path of Li atoms in plasma declines significantly. Thus they do not reach a substrate, leading to the formation of the lithium deficit LiNb$_3$O$_8$ phase. Consequently, this results in the generation of Li vacancies.

Figure 2.22 demonstrates the morphology of the films, fabricated by RFMS in an Ar atmosphere at high reactive gas pressure.

Analysis of the AFM images indicates that surface morphology of the films changes greatly when the working pressure increases in a reactive chamber. The average surface roughness of the films synthesized in this regime is around 60 nm, which is almost three times higher than those for the films deposited at lower working pressure (regime 2 in table 2.4). Furthermore, as follows from the study of

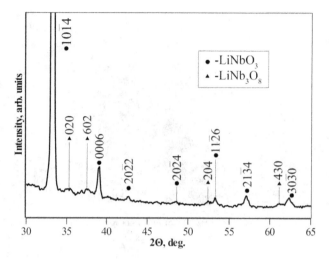

Figure 2.21. XRD pattern of the films, deposited on the (001)Si substrate by RFMS method under regime 4 (see table 2.4).

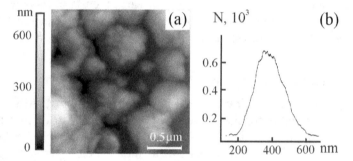

Figure 2.22. AFM images obtained during topography modes (a) and height distribution curves, showing surface roughness (b) of LiNbO₃ films, deposited on (001)Si substrates by RFMS method under condition 4 (see table 2.4).

elemental composition (RBS data) given in figure 2.23, an intermediated layer with a thickness of about 0.5 μm is formed at the Si/LiNbO$_3$ interface. Apparently, high working pressure in a reactive chamber leads to not only the decline in concentration of Li atoms in plasma, but also causes an intensive bombardment of the film surface, forming a defect layer. This layer facilitates inter-diffusion of atoms and interface reactions, forming the intermediate layer.

With the aim to study how reactive plasma influences films' structure at the initial growth stages we have synthesized thin films with a thickness of 100 nm on the (001)Si substrates by the RFMS method according to regimes 1, 2 and 3 in table 2.4 varying the substrate location in a horizontal plane according to figure 2.24. Figure 2.24 presents results of TEM study for as-grown films.

Taking into account that the reactive gas pressure over the target erosion zone (regime 1 in table 2.4, position 1 in figure 2.24(a)) is higher than those when a substrate is situated offset to the target erosion zone (regimes 2 and 3 in table 2.4)

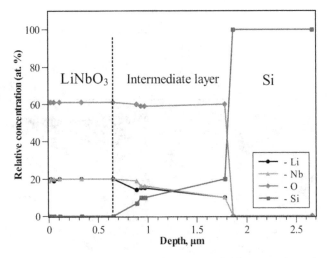

Figure 2.23. RBS spectrum for the film, synthesized on the (001)Si substrate under regime 4 (see table 2.4).

due to spatial plasma inhomogeneity, TEM patterns, shown in figure 2.24 (b)–(g) correspond to the films, deposited when reactive gas pressure rises. It follows from the TEM patterns, when the reactive gas pressure is low (see figure 2.24(b) and (c)) polycrystalline films with a grain size of 70 nm are formed. The clear-cut character of the grain boundaries indicates that an equilibrium structure is formed. All reflexes in the microdiffraction pattern (see figure 2.24(c)) correspond to the rhomboedrical lattice of LiNbO₃. The fact that the $11\overline{2}0$ reflex has a relatively high intensity indicates the formation of textured <0001>LiNbO₃ films. When reactive gas pressure increases, the grain size decreases to 30 nm (figure 2.24 (b), (d), (f)) along with the disappearance of film texture. Moreover, in the second and third cases (figure 2.24 (e) and (g)) a halo is observed which is attributed to inelastic electron scattering. This indicates the formation of an amorphous LiNbO₃ layer due to the effect of bombardment of the film surface by plasma particles, which is consistent with results reported in [7] and [4]. Also, it is important to note that the films, fabricated at the highest reactive gas pressure contain two phases (besides the amorphous phase): LiNbO₃ and the oxygen deficit LiNbO₂ phase.

2.3.3 Effect of the substrate type

As was demonstrated earlier, Si substrates do not manifest an orienting effect on the deposited LiNbO₃ films due to the formation of a thin (5 nm) amorphous SiO₂ layer at the Si/LiNbO₃ interface. A similar effect was reported in [11], where a thin amorphous intermediate layer was observed during deposition of LiNbO₃ films onto Si substrates. However, it is still unclear why textured LiNbO₃ films are formed on an amorphous SiO₂ layer. Some investigators believe that the key factor influencing this phenomenon is that the <0001>-axis is the polarization axis for lithium niobate. It was stressed in [12] that the conditions for a directional growth of <0001>-textured LiNbO₃ films occur when a deposition is conducted in external electric

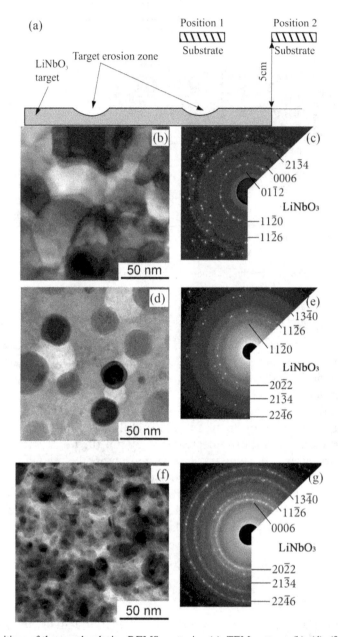

Figure 2.24. Positions of the samples during RFMS sputtering (a), TEM patterns (b), (d), (f) and the micro-diffraction pattern (c), (e), (g) of the film, formed at different reactive gas pressures (see comments in text) [4]. Reprinted by permission from Nature. Copyright 2015.

fields. Evidently, in our case, the intrinsic electric field, attributed to plasma in the RFMS method and directed perpendicular to a substrate surface, ensures directional growth of LiNbO$_3$ films along the field direction. Since atomic migrations along a surface in the framework of the Volmer–Weber growth mechanism are strongly affected by charged particles near this surface, it accounts for the decrease in crystallite size when working pressure increases. This is due to a strong dependence of the particle's energy on reactive gas pressure and substrate–target distance. A sufficient number of oxygen atoms, supplied by plasma to a substrate surface and also migration of atoms along the surface, activated by plasma, provides a 'stacking' sequence of lithium, niobium and oxygen atoms in two-dimensional O–Li–O–Nb–O layers (layer by layer) along the polar axis. At higher working pressure, intensive bombardment of a surface by ions occurs, inducing the formation of a disordered surface layer with a high concentration of vacancies. Numerous vacancies in an amorphous layer serve as centers, capturing Li atoms and intensifying reactions at the interface, which is enhanced by the capture of electrons by traps in this layer. At the same time, the bombardment leads to breaking of Si–O–Si bonds in a SiO$_2$ layer, enhancing the diffusion of Li atoms into a substrate and making out-of-diffusion of Si atoms toward the film surface possible. All of these factors create conditions for the formation of an intermediate layer of a complex composition and the LiNbO$_2$ phase at the substrate/film interface, which is observed in our case at high reactive gas pressure.

As regards the epitaxial (111)Ag substrates, a very strong orienting effect occurs because under this regime two-axis (epitaxial) texture is formed with the orientation ratio of (0001), [11$\bar{2}$0] LiNbO$_3$ ∥ (111), <110> Ag, providing the mosaic substructure of fabricated LiNbO$_3$ films. The mechanism of this phenomenon is not clear and is a question to be studied separately.

2.3.4 Effect of oxygen presence in the reactive gas environment

Based on the review in chapter 1, we saw that the presence of oxygen in a reactive gas chamber positively affects the structure and composition of deposited films. Nevertheless, the effect of the introduction of oxygen atoms in a reactive chamber can be different due to the mutual influence of various technological parameters of the RFMS method in every particular regime.

To study how the presence of oxygen in a reactive plasma influences the properties of synthesized LiNbO$_3$-based heterostructures we deposited the films by the RFMS method in an Ar + O$_2$ plasma. The oxygen content was chosen to be Ar/O$_2$ = 60/40 according to recommendations discussed in chapter 1. For comparison we also synthesized the films in the same sputtering regimes but in a pure Ar environment.

The films with a thickness of 0.5 μm were fabricated by the RFMS technique in an Ar(60%) + O$_2$(60%) atmosphere at a magnetron power of 100 W. The reactive gas pressure of $P = 1.5 \times 10^{-1}$ Pa was chosen because at this pressure highly oriented single-phase LiNbO$_3$ films are formed under these conditions, based on the experimental results discussed earlier. In the process of RFMS, substrates were situated offset to the target erosion zone but under the ion assist effect. The film's growth under the optimal working pressure of 1.5×10^{-1} Pa with the ion assist effect is accompanied by a higher degree of flux laminarity which results in the formation of a single-axis texture. The film's growth rate under these conditions at a distance of 5 cm between the heated (001)Si substrate ($T = 550\ ^\circ$C) and a target was 10 nm min^{-1}.

Figure 2.25 shows the comparative results of the study of elemental composition of the films, deposited in an Ar + O$_2$ gas mixture and in a pure Ar atmosphere.

As follows from figure 2.25, elemental composition of the films fabricated in both environments corresponds to stoichiometric LiNbO$_3$. On the other hand an intermediated SiO$_2$ layer with a thickness of about 0.6 μm is formed when films are deposited in an Ar + O$_2$ reactive gas mixture. This layer grows inevitably due to the presence of oxygen in reactive plasma, which results in oxidation of the Si substrate surface with formation of SiO$_2$ layer.

The XRD pattern, shown in figure 2.26, indicates that highly oriented single-axis <0001>-textured LiNbO$_3$ films are formed during RFMS process in an Ar + O$_2$ reactive gas atmosphere. This fact is in a good agreement with other authors [13, 14] and with our results [15].

According to calculations using the Selyakov–Scherrer method, an average grain size is around 40 nm, which is close to those in the films deposited in a pure Ar atmosphere.

Figures 2.27 and 2.28 are AFM images of LiNbO$_3$ films, fabricated in an Ar atmosphere and in an Ar + O$_2$ gas mixture, respectively.

The surface relief consists of elongated inhomogeneities with the following dimensions: width up to 50 nm, length up to 150 nm. The average surface roughness is 10 nm for the scanned area of 2×2 μm^2 (see figure 2.27). The surface relief of the films, grown in an Ar + O$_2$ atmosphere manifests inhomogeneities with a lateral dimension of 60 nm (see figure 2.28). The average surface roughness is 3 nm, which is considerably lower than for the films fabricated in an Ar environment. The lateral dimensions of the studied films correlate with an average size of the coherent scattering region, calculated from XRD patterns. Low surface roughness is a crucial

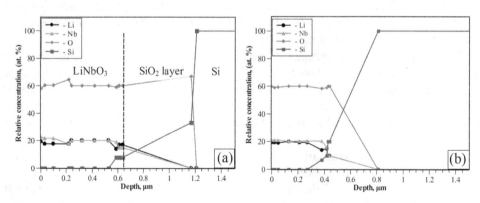

Figure 2.25. RBS spectrum for (001)Si–LiNbO$_3$ heterostructures grown in an Ar + O$_2$ gas mixture (a) and in an Ar gas environment (b).

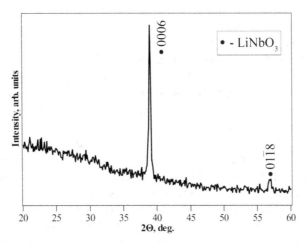

Figure 2.26. XRD patterns of the films with a thickness of about 1.0 μm, fabricated in an Ar + O$_2$ environment.

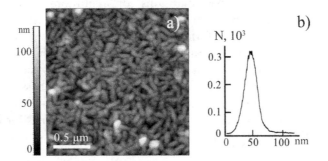

Figure 2.27. AFM images obtained during topography modes (a) and height distribution curves, showing surface roughness (b) of LiNbO$_3$ films, fabricated on (001)Si substrate by the RFMS method in an Ar environment. Reprinted from [15]. Copyright (2014), with permission from Elsevier.

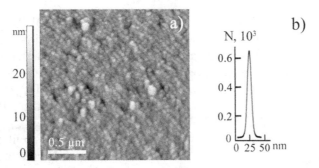

Figure 2.28. AFM images obtained during topography modes (a) and height distribution curves, showing surface roughness (b) of LiNbO$_3$ films, deposited by the RFMS method on (001)Si substrate in an Ar + O$_2$ environment. Reprinted from [15]. Copyright (2014), with permission from Elsevier.

parameter for electrical contacts, which makes the RFMS in an Ar + O_2 gas mixture more efficient from this point of view.

In our opinion the formation of a single-axis texture is caused by the plasma effect (the ion assist effect) as well as the presence of oxygen in a reactive chamber (to a lesser degree). To support this opinion some authors note an extremely important role which oxygen plays in formation of oriented $LiNbO_3$ films. For example, it is noted in [16] that $LiNbO_3$ films, fabricated in an Ar atmosphere on atomically clean Si surfaces without a natural oxide, have an arbitrary grain orientation. On the other hand, when oxygen presents in a reactive chamber it covers a growing surface with a two-dimensional atomic layer. Atomic migration, activated by plasma, forces the formation of such a layer and consequently, a single-axis texture. The extended SiO_2 layer prevents formation of the texture. As follows from [16] Li atoms can react with the SiO_2 film, forming a double-atomic layer, the top of which is the LiNbSiO compound. This top layer blocks the formation of the $LiNbO_3$ phase. This agrees with our results, described above, when only $LiNbO_3$ films with a random grain orientation are formed on the Si–SiO_2 heterostructures. Apparently, the formation of oxygen atomic mono-layers on the surface creates favorable conditions for migration of adatoms, creating minimal potential barriers between islands of a new phase at the beginning of the film growth. It facilitates the formation of a smoother surface, compared to the films, grown in a pure Ar environment. To support this opinion we refer to the results of [13], where a strong dependence of surface roughness on oxygen pressure in a reactive chamber is demonstrated. Minimal surface roughness of 9 nm given in this paper corresponds to the oxygen pressure of 40 Pa.

Thus, based on our study the most optimal sputtering regimes for fabrication of single-phase $LiNbO_3$ films are given in table 2.5.

2.3.5 Effect of thermal annealing

Thermal annealing (TA) in an air or oxygen atmosphere is one of the most effective post-growing treatments of as-grown films. Two processes play an important role: re-crystallization and diffusion of oxygen with further reactions in bulk and at the interfaces.

To study the effect of TA we have fabricated films on the (001)Si substrates under the ion assist effect offset from the target erosion zone. Deposition regimes are given in table 2.6. The annealing temperature of 600 °C has been chosen because at this temperature evaporation of components begins from the surface of $LiNbO_3$ films, which allows influence of their properties effectively through TA [8].

Figure 2.29 presents the results of composition study by the RBS method for samples LN1-T, LN_2-T and LN3-T.

As follows from comparison between RBS spectra of as-grown films (figures 2.23 and 2.25) and films after TA (figure 2.29), thermal annealing in an air environment does not influence significantly elemental composition of the studied films.

Figure 2.30 demonstrates XRD patterns of the films, deposited under regimes, given in table 2.5 after TA.

Table 2.5. The optimal sputtering regimes for thin LiNbO$_3$ films, fabricated by IBS and RFMS methods[‡].

Deposition method	Substrate	Reactive gas environment	Magnetron (supply) power	Substrate temperature	Reactive gas pressure (Pa)	Substrate-target distance and position (cm)	Properties of LiNbO$_3$ films
RFMS	Si	Ar	100 W	550 °C	0.15	4–5 (offset of the target erosion zone)	Single-phase <0001>-textured LiNbO$_3$ films. Average grain size is 40 nm. Average surface roughness is 10 nm.
	Si	Ar				8–10 (offset of the target erosion zone)	Single-phase LiNbO$_3$ films with random grain orientation. Average grain size is 50 nm. Average surface roughness is 16 nm.
	Si	Ar + O$_2$ (60/40)				4–5 (offset of the target erosion zone)	Single-phase <0001>-textured LiNbO$_3$ films with minimal surface roughness (3 nm). Average grain size is 40 nm.
	(111)Ag					4–5 (offset of the target erosion zone)	Single-phase LiNbO$_3$ films with epitaxial texture with the orientation ratio of (0001), [11$\bar{2}$0] LiNbO$_3$ ∥ (111), <110> Ag.
IBS	Si	Ar	2 kW			4–5 (offset of the target erosion zone)	Single-phase LiNbO$_3$ films with random grain orientation. Average grain size is 50 nm. Average surface roughness is 40 nm.

[‡]Shaded cells correspond to the films, fabricated without the ion assist effect.

Table 2.6. Synthesis regimes for samples which have undergone TA treatment.

Sample #	Reactive gas composition	Reactive gas pressure (Pa)	TA at temperature of 600 °C	Substrate-target distance and position (cm)
LN1	Ar	0.5	-	5
LN1-T			+	
LN2		0.15	-	
LN2-T			+	
LN3	Ar + O_2	0.15	-	
LN3-T			+	

As follows from figure 2.30, compared to as-grown films (see figures 2.18, 2.21 and 2.26), TA results in re-crystallization of the films with a two times increase (up to 80–100 nm) in average grain size. Also, it is important to stress, that TA leads to the formation of the Li-poor $LiNb_3O_8$ phase in the films, synthesized at reactive gas pressure of 0.15 Pa (see samples LN2-T and LN3-T). Furthermore, in as-grown films, having <0001>-texture (samples LN1 and LN2), TA results in the disappearance of this texture with the formation of polycrystalline films with a random grain orientation. This experimental fact is in good agreement with the data of numerous investigators [17–21] who have observed the formation of this 'parasitic' phase after TA both in vacuum and in oxygen. As was noted in chapter 1 this phenomenon is caused by the loss of lithium and oxygen by a film due to desorption of Li_2O during thermal annealing.

The results of morphological study of the films after TA are given in figures 2.31–2.33 [22].

Analysis of AFM images, shown in figures 2.31–2.33 indicates twofold increase of an average surface roughness of the studied films after TA. The average surface roughness is 70 nm, 14 m and 15 nm for samples LN1-T, LN2-T and LN3-T, respectively. The most pronounced increase was observed for the films fabricated in an Ar + O_2 gas mixture: from 3 nm for as-grown films (see figure 2.28 for sample LN3) to 14 nm for the films after TA (see figure 2.33). It is worth noting the increase of relief inhomogeneities after TA that are 300 nm, 200 nm and 150 nm for samples LN1-T, LN2-T and LN3-T, respectively [22]. This size agrees with the size estimated from XRD patterns.

As regards $LiNbO_3$ films fabricated by the IBS method, it follows from the TEM patterns shown in figure 2.34, the TA of as-grown films results in the formation of the lithium triniobate phase ($LiNb_3O_8$) and an increase of average grain size from 50 nm to 100 nm [23].

As was revealed for the films synthesized by RFMS technique, $LiNbO_3$ films fabricated by the IBS method manifest an arbitrary grain orientation after TA along with an increase of surface roughness from 50 nm to 100 nm [24].

To sum up, it is important to note some general trends, attributed to all $LiNbO_3$-based heterostructures after TA regardless of the deposition method. Firstly, the average grain size and surface roughness increases after TA along with the

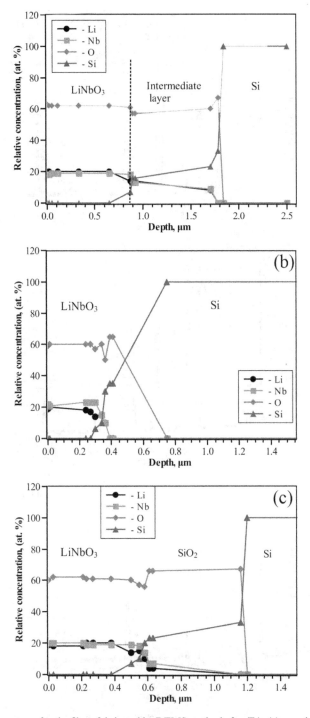

Figure 2.29. RBS spectrum for the films, fabricated by RFMS method after TA. (a) sample LN1-T, (b) sample LN2-T, (c) sample LN3-T.

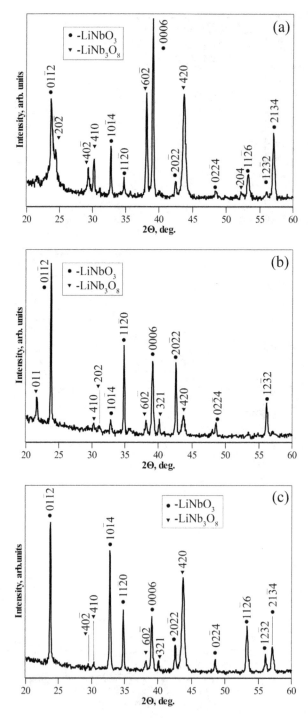

Figure 2.30. XRD patterns of the samples LN1-T (a), LN2-T (b) and LN3-T (c) (see table 2.6).

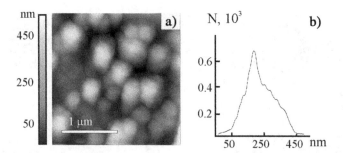

Figure 2.31. AFM images obtained during topography modes (a) and height distribution curves showing surface roughness (b) of $LiNbO_3$ films, related to the sample LN1-T [22]. Reprinted by permission from Springer Nature. Copyright 2015.

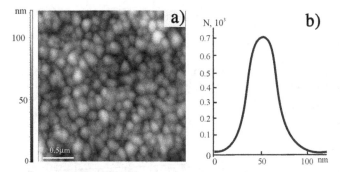

Figure 2.32. AFM images obtained during topography modes (a) and height distribution curves showing surface roughness (b) of $LiNbO_3$ films related to the sample LN2-T [22]. Reprinted by permission from Springer Nature. Copyright 2015.

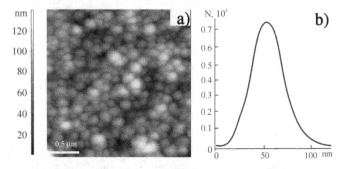

Figure 2.33. AFM images obtained during topography modes (a) and height distribution curves showing surface roughness (b) of $LiNbO_3$ films, related to the sample LN3-T [22]. Reprinted by permission from Springer Nature. Copyright 2015.

disappearance of texture in as-grown $LiNbO_3$ films. Second, the 'parasitic' $LiNb_3O_8$ phase is formed in the films after TA, which is a consequence of desorption of volatile compounds such as Li_2O. Some authors argue that the $LiNb_3O_8$ phase is formed due to the separation of the $Li_{1-x}Nb_{1+y}O_3$ phase into two rival phases $LiNbO_3$ and $LiNb_3O_8$ [25]. In contrast, others believe that the $LiNb_3O_8$ phase is

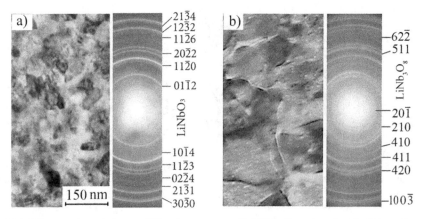

Figure 2.34. Micrographs and electron diffraction patterns of LiNbO$_3$ films, deposited by IBS method on the (001)Si substartes. (a) as-grown films, (b) films after TA. Reprinted from [23]. Copyright (2011), with permission from Society of Photo-Optical Instrumentation Engineers (SPIE).

formed on a surface due to desorption of LiO$_2$ because of the high concentration of oxygen vacancies which occur near the surface, promoting separation of LiNbO$_3$ and LiNb$_3$O$_8$ phases [21].

Summary and emphasis

1. The possibility of fabrication of single-phase LiNbO$_3$ films by RFMS and IBS methods in Ar and in Ar + O$_2$ reactive environments has been demonstrated. The RFMS method has more advantages compared to the IBS method due to higher flexibility of its sputtering regimes, higher growth rate and the possibility of fabrication of highly oriented films on various substrates.

2. The RFMS method possesses a range of critical parameters greatly influencing structure, composition and surface morphology. The following patterns have been revealed:

   - *Plasma effect (the ion assist effect).* When an appropriate relative position of a substrate and a target–substrate distance are chosen, plasma particles cause a directional growth of LiNbO$_3$ film on all types of substrates, influencing a deposition process. The <0001>-texture is formed on (001)Si, (111)Si wafers and on the Si–SiO$_2$ heterostructures. From this point of view the optimal substrate position is 'offset the target erosion zone' at a distance of 4–5 cm under working pressure and power of 0.15 Pa and 100 W, respectively.

   - *Effect of reactive gas pressure.* When reactive gas pressure increases in a reactive chamber, only films with a random grain orientation are formed, and grain size decreases with pressure. This is caused by heavy bombardment of a surface by plasma particles, making it amorphous at the beginning growth stages with the formation of an intermediate layer at the substrate/film interface. A strong bombardment of the film's surface and the decline in the mean free path of plasma particles lead to a decrease in the concentration of Li atoms near the growing surface. The $LiNb_3O_8$ phase is formed along with $LiNbO_3$ in the films.

- *Types of substrates.* Films deposited onto the (001)Si and (111)Si wafers, do not manifest the orienting substrate effect due to formation of a thin (5 nm) buffer SiO_2 layer, passivating the substrates. In contrast, epitaxial $LiNbO_3$ films are fabricated on the (111)Ag substrates by the RFMS method. The two-axis (epitaxial) texture is formed with the orientation ratio of (0001), $[11\bar{2}0]$ $LiNbO_3$ ‖ (111), <110> Ag, ensuring the mosaic substructure of fabricated $LiNbO_3$ films. Thus, it is advisable to use the epitaxial Ag films as substrates for deposition of $LiNbO_3$ films (and specifically for the synthesis of the films separated from a substrate).

3. The presence of oxygen along with Ar in a reactive chamber results in the formation of highly oriented <0001>-$LiNbO_3$ films with minimal surface roughness. The extended SiO_2 layer is formed at the substrate/film interface due to oxidation of a surface in the presence of oxygen.

4. Thermal annealing of as-grown textured $LiNbO_3$ films in air leads to their recrystallization, increasing the average surface roughness and grain size twofold (up to 15–70 nm and 80–100 nm, respectively). TA results in the formation of the $LiNb_3O_8$ phase in films, fabricated under working pressure of 1.5×10^{-1}Pa. The texture existing in as-grown films is leveled and the grain orientation becomes arbitrary.

References

[1] Iyevlev V, Kostyuchenko A, Sumets M and Vakhtel V 2011 Electrical and structural properties of $LiNbO_3$ films, grown by RF magnetron sputtering *J. Mater. Sci. Mater. Electron.* **22** 1258–63

[2] Simões A Z, Zaghete M A, Stojanovic B D, Riccardi C S, Ries A, Gonzalez A H and Varela J A 2003 $LiNbO_3$ thin films prepared through polymeric precursor method *Mater. Lett.* **57** 2333–9

[3] Rabson T A, Baumann R C and Rost T A 1990 Thin film lithium niobate on silicon *Ferroelectrics* **112** 265–71

[4] Sumets M, Kostyuchenko A, Ievlev V, Kannykin S and Dybov V 2015 Sputtering condition effect on structure and properties of $LiNbO_3$ films *J. Mater. Sci. Mater. Electron.* **26** 4250–6

[5] Sumets M, Dybov V A and Ievlev V M 2017 $LiNbO_3$ films: potential application, synthesis techniques, structure, properties Inorg. Mater.**53** 1361–77

[6] Ievlev V, Sumets M and Kostyuchenko A 2013 Conduction mechanisms in Si-$LiNbO_3$ heterostructures grown by ion-beam sputtering method *J. Mater. Sci.* **48** 1562–70

[7] Rost T A, Lin H, Rabson T A, Baumann R C and Callahan D L 1992 Deposition and analysis of lithium niobate and other lithium niobium oxides by rf magnetron sputtering *J. Appl. Phys.* **72** 4336–43

[8] Park S K, Baek M S, Bae S C, Kwun S Y, Kim K T and Kim K W 1999 Properties of $LiNbO_3$ thin film prepared from ceramic Li-Nb-K-O target *Solid State Commun.* **111** 347–52

[9] Kong Y, Xu J, Chen X, Zhang C, Zhang W and Zhang G 2000 Ilmenite-like stacking defect in nonstoichiometric lithium niobate crystals investigated by Raman scattering spectra *J. Appl. Phys.* **87** 4410

[10] Blümel J, Born E and Metzger T 1994 Solid state NMR study supporting the lithium vacancy defect model in congruent lithium niobate *J. Phys. Chem. Solids* **55** 589–93

[11] Akazawa H and Shimada M 2004 Correlation between interfacial structure and c-axis-orientation of $LiNbO_3$ films grown on Si and SiO_2 by electron cyclotron resonance plasma sputtering *J. Cryst. Growth* **270** 560–7

[12] Hu W S, Liu Z G, Lu Y-Q, Zhu S N and Feng D 1996 Pulsed-laser deposition and optical properties of completely (001) textured optical waveguiding $LiNbO_3$ films upon SiO_2/Si substrates *Opt. Lett.* **21** 946

[13] Wang X, Liang Y, Tian S, Man W and Jia J 2013 Oxygen pressure dependent growth of pulsed laser deposited $LiNbO_3$ films on diamond for surface acoustic wave device application *J. Cryst. Growth* **375** 73–7

[14] Tan S, Gilbert T, Hung C-Y, Schlesinger T E and Migliuolo M 1996 Sputter deposited c-oriented $LiNbO_3$ thin films on SiO_2 *J. Appl. Phys.* **79** 3548

[15] Sumets M, Ievlev V, Kostyuchenko A, Vakhtel V, Kannykin S and Kobzev A 2014 Electrical properties of Si-$LiNbO_3$ heterostructures grown by radio-frequency magnetron sputtering in an Ar + O_2 environment *Thin Solid Films* **552** 32–8

[16] Akazawa H and Shimada M 2006 Factors driving c-axis orientation and disorientation of $LiNbO_3$ thin films deposited on TiN and indium tin oxide by electron cyclotron resonance plasma sputtering *J. Appl. Phys.* **99** 124103

[17] Simões A Z, Zaghete M A, Stojanovic B D, Gonzalez A H, Riccardi C S, Cantoni M and Varela J A 2004 Influence of oxygen atmosphere on crystallization and properties of $LiNbO_3$ thin films *J. Eur. Ceram. Soc.* **24** 1607–13

[18] Kiselev D A, Zhukov R N, Bykov A S, Voronova M I, Shcherbachev K D, Malinkovich M D and Parkhomenko Y N 2014 Effect of annealing on the structure and phase composition of thin electro-optical lithium niobate films *Inorg. Mater.* **50** 419–22

[19] Bornand V, Huet I and Papet P 2002 $LiNbO_3$ thin films deposited on Si substrates: a morphological development study *Mater. Chem. Phys.* **77** 571–7

[20] Akazawa H and Shimada M 2006 Precipitation kinetics of $LiNbO_3$ and $LiNb_3O_8$ crystalline phases in thermally annealed amorphous $LiNbO_3$ thin films *Phys. status solidi* **203** 2823–7

[21] Akazawa H and Shimada M 2007 Mechanism for $LiNb_3O_8$ phase formation during thermal annealing of crystalline and amorphous $LiNbO_3$ thin films *J. Mater. Res.* **22** 1726–36

[22] Sumets M, Kostyuchenko A, Ievlev V, Kannykin S and Dybov V 2015 Influence of thermal annealing on structural properties and oxide charge of $LiNbO_3$ films *J. Mater. Sci. Mater. Electron.* **26** 7853–9

[23] Iyevlev V, Kostyuchenko A and Sumets M 2011 Fabricatoin, substructure and properties of $LiNbO_3$ films *Proc. SPIE* **7747** 77471J

[24] Ievlev V M, Sumets M P and Kostyuchenko A V 2012 Effect of thermal annealing on electrical properties of Si-$LiNbO_3$ *Mater. Sci. Forum* **700** 53–7

[25] Esdaile R J 1985 Comment on "Characterization of TiO_2, $LiNb_3 O_8$, and $(Ti_{0.65} Nb_{0.35})O_2$ compound growth observed during Ti:$LiNbO_3$ optical waveguide fabrication" *J. Appl. Phys.* **58** 1070–1

Chapter 3

Electron phenomena in LiNbO$_3$-based heterostructures

3.1 Basic electrical properties of LiNbO$_3$ thin films in Si–LiNbO$_3$ heterosystems

As demonstrated in chapter 1, the electrical properties of thin LiNbO$_3$ films (dielectric constant, conductivity, band gap, remnant polarization etc) influence the functionality of integrated electronic and optoelectronic devices. In this chapter we study the main electrical properties of LiNbO$_3$ films and Si–LiNbO$_3$ heterostructures. We have fabricated (001)Si–LiNbO$_3$ and (001)Si–SiO$_2$–LiNbO$_3$ heterostructures by the RFMS and IBS methods, in the absence of the ion assist effect, according to the optimal regimes proposed in chapter 2. Therefore, we have to note that all results reported in chapter 3 are related to the heterostructures fabricated *without the ion assist effect*. Silicon wafers ((001)Si n- and p-type conductivity with $\rho = 20$ Ohm cm and $\rho = 4.5$ Ohm cm, respectively) and the Si–SiO$_2$ heterostructures were used as substrates. Si–SiO$_2$ heterostructures were fabricated by the annealing of Si wafers in a coaxial furnace at 700 °C. In the framework of the RFMS method, deposition was performed on substrates, heated to 550 °C and located offset from the target erosion zone (8–10 cm apart), which eliminates the ion assist effect. Analysis of the composition and structure of thin LiNbO$_3$ films, discussed in chapter 2, suggests that in these regimes single phase polycrystalline LiNbO$_3$ films with arbitrary grain orientation are formed. Electrical properties of fabricated heterostructures were studied using techniques based on obtaining the current–voltage (I–V) and high frequency (1 MHz) capacitance–voltage (C–V) characteristics, and also based on the tangent loss frequency dependence and impedance spectroscopy in the frequency range of 30–10^4 Hz with temperatures ranging from 77 K to 400 K. I–V characteristics were measured at the time rate of voltage change of $dV/dt = 0.1$ V s^{-1}. The ferroelectric properties were studied by recording the hysteresis loops using the Sawyer–Tower method. The top contacts for the electrical measurements had an area of $S = 1 \times 10^{-6}$ m^2 and were formed by thermal evaporation and condensation of Al in vacuum (1×10^{-4} Pa). The bottom electrode was created using an In/Ga eutectic alloy on the Si substrate that provides the formation of the Ohmic contacts [1].

3.1.1 Capacitance–voltage and current–voltage characteristics of LiNbO$_3$-based heterostructures

Typical high frequency ($f = 1$ MHz) C–V characteristics of (001)Si–SiO$_2$–LiNbO$_3$–Al heterostructures, fabricated by the RFMS method on Si substrates of p-type conductivity, accorded to C–V characteristics of metal–insulator–semiconductor (MIS) systems is shown in figure 3.1. Analysis of C–V characteristics was conducted according to the standard methods of C–V spectroscopy [2]. The studied (001)Si–SiO$_2$–LiNbO$_3$–Al heterostructures can be represented as MIS structures with double layered dielectric SiO$_2$/LiNbO$_3$. In this case its capacitance in the accumulation regime C_i is equivalent to the capacitance of capacitors C_{LN} and C_{ox} connected in series and representing the capacitance of LiNbO$_3$ and SiO$_2$ layers, respectively.

Using the formulas for a parallel-plate capacitor and for capacitance of two capacitors in series we have the following system of equations:

$$C_{\mathrm{LN}} = \frac{\varepsilon_{\mathrm{LN}}\varepsilon_0 S}{d}$$

$$C_{\mathrm{ox}} = \frac{\varepsilon_{\mathrm{ox}}\varepsilon_0 S}{d_{\mathrm{ox}}}$$

$$\frac{1}{C_i} = \frac{1}{C_{\mathrm{LN}}} + \frac{1}{C_{\mathrm{ox}}}$$

(3.1)

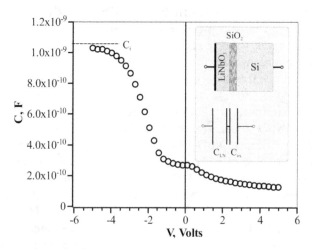

Figure 3.1. Typical high frequency (1 MHz) C–V characteristics of a (001)Si–SiO$_2$–LiNbO$_3$–Al heterostructure. The inset shows schematically the studied heterostructure and its equivalent circuit in the accumulation regime, when measured capacitance is C_i.

Here C_i is the capacitance of an MIS capacitor in the accumulation regime, S is the contact area, $\varepsilon_0 = 8.85 \times 10^{-12}$ F m^{-1} is the electric constant, ε_{LN} and ε_{ox} are the dielectric constants of LiNbO$_3$ and SiO$_2$, respectively, d and d_{ox} are the thickness of the LiNbO$_3$ film and a silicon dioxide layer. Solving equation (3.1) for ε_{LN} we obtain:

$$\varepsilon = \varepsilon_{ox} \frac{C_i}{\varepsilon_{ox}\varepsilon_0 S/d - C_i d_{ox}/d} \tag{3.2}$$

Taking into account that for silicon dioxide $\varepsilon_{ox} = 3.82$ and for the studied heterostructures $d_{ox} = 120$ nm, we obtain the dielectric constant of the LiNbO$_3$ films from equation (3.2): $\varepsilon_{LN} = 28$. This result is in good agreement with that reported in [3, 4] for thin LiNbO$_3$ films and close to the dielectric permittivity of bulk lithium niobate ($\varepsilon = 30$ [5]). The fact that the $C–V$ curve is shifted to the left along the voltage axis (see figure 3.1) is evidently attributed to the presence of the positive fixed charge in the films. The energy distribution of surface states at the Si–SiO$_2$ interface in the lower half of the Si bang gap is calculated according to [2] through the shift of an experimental $C–V$ curve relative to the theoretical one (see figure 3.2).

Analysis of $C–V$ characteristics reveals that the effective density of interface states is $N_{eff} = 2.4 \times 10^{11}$ cm^{-2}. Also, it follows from figure 3.1 that at zero bias the studied heterostructures are in the deep depletion regime close to the inversion regime, where a strong inversion layer is formed at a dielectric–semiconductor interface, which is caused by a strong internal electric field in a dielectric. At this condition even low direct bias ('−' at the metal electrode) can be sufficient to generate leakage currents, that restricts application of synthesized heterostructures as nonvolatile memory units and optoelectronic devices.

High frequency $C–V$ characteristics of (001)Si–SiO$_2$–LiNbO$_3$–Al heterostructures, fabricated under the same regimes onto the p-type (001)Si wafers (without the SiO$_2$ layer) were also similar to the $C–V$ characteristics of the MIS capacitor having high density of interface states, which prevents its saturation (see figure 3.3).

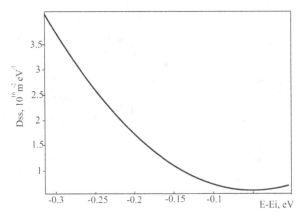

Figure 3.2. Energy distribution of surface states at the Si–SiO$_2$ interface in the (001)Si–SiO$_2$–LiNbO$_3$–Al heterostructure.

At higher negative bias the leakage current in (001)Si–LiNbO$_3$–Al heterostructures, fabricated on p-Si, prevents the use of the C–V analysis properly, so we use I–V methods for this heterostructure.

I–V characteristics of (001)Si–LiNbO$_3$–Al heterostructures, fabricated on p-Si, measured at $T = 300$ K in the ln J–V coordinates are shown in figure 3.4.

Resistivity of fabricated LiNbO$_3$ films is determined from an initial (Ohmic) section of I–V curves and is equal to $\rho = 1 \times 10^9$ Ohm cm for the films, deposited onto (001)Si substrates, which is in a good agreement with the results of [6].

Considering the possible application of the diffusion theory for the analysis of I–V characteristics, we estimate the Debye length $L_D = \sqrt{\varepsilon\varepsilon_0 kT/q^2 n_0}$ and the mean free path $l = V_T\langle\tau\rangle = V_T m\mu/q$ (here ε and ε_0 are the dielectric permittivity of a material and vacuum, respectively, k is Boltzmann's constant, q and m are the electronic charge and mass, n_0 is the concentration of free charges, V_T, μ and $\langle\tau\rangle$ are the

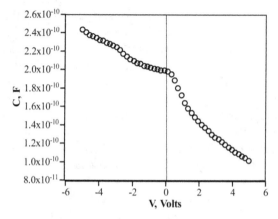

Figure 3.3. Typical high frequency (1 MHz) C–V characteristics of the (001)Si–LiNbO$_3$–Al heterostructure, fabricated by the RFMS method on p-Si.

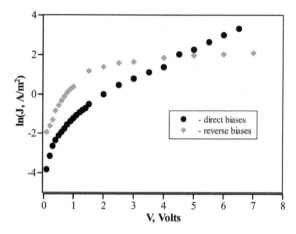

Figure 3.4. Typical I–V characteristics of p-(001)Si–LiNbO$_3$–Al heterostructures, fabricated by RFMS at direct bias ('−' at Al contact) and reverse bias ('+' at Al contact).

thermal velocity, mobility and the average effective relaxation time of carriers). In our case for Si ($n_0 \sim 10^{15}$ cm^{-3}, $\mu \approx 0.5 \times 10^3$ cm^2 V^{-1}s^{-1}, $T = 300$ K, $\varepsilon = 12$, $V_T \approx 1 \times 10^7$cm s^{-1}) the Debye length and mean free path have the magnitudes of $L_D \sim 10^{-5}$ cm, and $l \sim 3 \times 10^{-2}$ cm, respectively. Because $l \gg L_D$, the diode theory should be applied whereas the diffusion currents are negligible.

I–V characteristics of the studied heterostructures can be described in the framework of the model of MIS structures obeying the following expression [7]:

$$J = J_s\left(\exp\left(\frac{qV}{n_1 \cdot kT}\right) - \exp\left(-\frac{qV}{n_2 \cdot kT}\right)\right) \qquad (3.3)$$

Here J is the current density, q is the electron charge, V is the applied voltage, k is the Boltzmann's constant, T is the temperature, n_1 and n_2 are the ideality factors, depending on the charge transport mechanism and J_s is the saturation current density, which in the thermionic emission-diffusion theory has the following form [8]:

$$J_s = A^* T^2 \exp\left(-\frac{qE_t}{kT}\right) \qquad (3.4)$$

Here A^* is the effective Richardson constant, described by the following expression [9]:

$$A^* = \frac{2q\mu}{d \cdot S} V\left(\frac{2\pi m^* k}{h^2 T^3}\right)^{1/2} \qquad (3.5)$$

where d is the thickness of a dielectric, S the contact area, E_t the energy of a monoenergetic level in the band gap of the dielectric, m^* the carrier's effective mass, h Planck's constant and μ is the carrier's mobility.

According to equation (3.3) linear extrapolation of the I–V characteristic in $\ln(J/T^2) - 1/T$ coordinates to $1/T \to 0$ gives the effective Richardson constant A^*, whereas the slope of this linear function allows the activation energy of conductivity to be determined.

The ideality factor is of great technical interest and it can be determined by graphical differential of a linear section of I–V curve in $\ln J - V$ coordinates using the following formula:

$$n = \frac{q}{kT}\frac{dV}{d(\ln J)} \qquad (3.6)$$

Direct biases
Usually, in the band gap and at the dielectric–semiconductor interface there are energy levels attributed to defects and mismatch of crystal lattices of two materials. For the charge transport over monoenergetic levels in a dielectric, the ideality factor can be approximately written as [8]:

$$n = \frac{d}{l_t} \qquad (3.7)$$

Here d is the thickness of a dielectric, $l_t = N_t^{-1/3}$ the average separation between trap centers in the dielectric and N_t is their concentration. Thus, the trap concentration in the band gap of a dielectric can be estimated through the ideality factor, obtained from the experimental I–V characteristic.

We have determined the effective Richardson constant and activation energy of conductivity from temperature dependence of current density for the studied heterostructures. The typical temperature dependence of conductivity in the Arrhenius coordinates is shown in figure 3.5.

Two temperature intervals associated with different activation processes correspond to two linear sections with different slopes in figure 3.5. The activation energy for each process has magnitudes of $E_{a1} = 0.23$ eV and $E_{a2} = 0.05$ eV.

By extrapolating the linear part of the I–V curve in $\ln J - V$ coordinates to $V\rightarrow 0$ (see figure 3.4), we obtain J_s and then, using equation (3.4) we determined the energy of traps (below the bottom of conduction band) E_t. The results, obtained for LiNbO$_3$ film with thickness of $d = 1 \times 10^{-6}$ m and dielectric permittivity $\varepsilon = 28$, are given in table 3.1.

Figure 3.6 shows experimental and theoretical I–V characteristics, calculated using equation (3.3) based on the parameters listed in table 3.1. As can be seen in

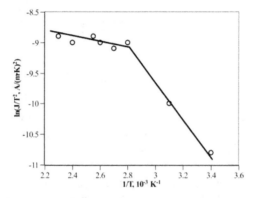

Figure 3.5. Typical temperature dependence of conductivity for (001)Si–LiNbO$_3$–Al heterostructures in $\ln (J/T^2) - 1/T$ coordinates at direct bias ('–' at Al contact).

Table 3.1. Results of analysis of I–V characteristics of p-(001)Si–LiNbO$_3$–Al heterostructures, fabricated by RFMS on p-Si substrates.[a]

Substrate	Resistivity of LiNbO$_3$ film, ρ (Ohm cm)	Ideality factor, n	Effective Richardson constant A^*, (A m^{-2} K^{-2})	Trap concentration in LiNbO$_3$, N_t (cm^{-3})	Energy position of traps in the band gap of LiNbO$_3$, E_t (eV)	Activation energy of conductivity E_a (eV)	The Poole–Frenkel coefficients ratio $\beta_{P-F}^{ex}/\beta_{P-F}^{theor}$
p-Si(001)	1×10^9	51(386)	0.1	2.4×10^{17} (2.4×10^{17})	0.27	0.23(0.1)	1.08

[a] Parameters, derived from the reverse branches of I–V characteristics, are indicated in brackets.

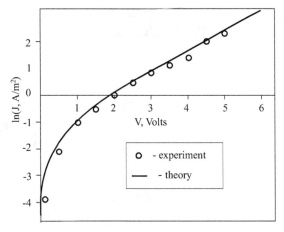

Figure 3.6. Experimental and theoretical I–V characteristics of (001)Si–LiNbO$_3$–Al heterostructures at the direct biases (dots—experiment, solid line—theoretical curve, calculated using equation (3.3)).

figure 3.6 the theoretical model described by equation (3.3), is in a good agreement with experimental data for (001)Si–LiNbO$_3$–Al heterostructures.

One of the possible conduction mechanisms over the centers of localized charge (CLC) can be the hopping conductivity where the temperature dependence of current density is defined by the following expression (the Mott's variable range hopping conductivity) [10]:

$$J(T) = \frac{J_0}{T^{1/2}} \exp(-(T_0/T)^{1/4}) \tag{3.8}$$

Here T_0 is a parameter, given by:

$$T_0 = \frac{\lambda}{kN(E_F)a^3} \tag{3.9}$$

where $N(E_F)$ is the energy density of localized states near the Fermi level, k is Boltzmann's constant, a is the localization radius, λ is a dimensionless parameter (usually $\lambda \sim 16$ [10]). In fact, we have demonstrated that the experimental I–V curve is perfectly linearized in Mott's coordinates [10] (figure 3.7), and the parameter T_0 can be obtained from the slope of this line.

In the framework of this model the average hopping distance R of carriers over the localized sites near the Fermi level at temperature T is defined by the following expression [11]:

$$R = \frac{3}{8}a\left(\frac{T_0}{T}\right)^{1/4} \tag{3.10}$$

The energy range of the localized states in this case is described by:

$$\Delta E = \frac{3}{2\pi N(E_F)R^3} \tag{3.11}$$

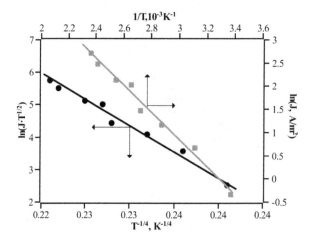

Figure 3.7. Temperature dependence of (001)Si–LiNbO₃–Al heterostructures conductivity in the Mott's and Arrhenius coordinates at a direct bias of $V = 2$ V.

Concentration of CLC can be defined as follows:

$$N_t = N(E_F)\Delta E \qquad (3.12)$$

For the studied heterostructures we have obtained the following results [12]: $\Delta E = 0.5$ eV, $N(E_F) = 5 \times 10^{17}$ eV^{-1} cm^{-3}, $N_t = 2.5 \times 10^{17}$ cm^{-3}.

Reverse biases
I–V characteristics at reverse biases should saturate according to the Schottky diode theory [2]. However, in our case this was not observed. This can be explained in the framework of the Poole–Frenkel effect (thermal ionization under electric fields), which is analogous to the Schottky emission, but manifested in the bulk of a material (see chapter 1). In this case, *I–V* characteristics can be described in terms of applied voltage by an expression similar to equation (1.15):

$$J = J_s \exp\left(\frac{\Delta\varphi}{kT}\right) \qquad (3.13)$$

here J_s is the saturation current and the decrease in potential barrier (the decrease in ionization energy of a single Coulomb potential well) is described by the Poole–Frenkel formula [13]:

$$\Delta\varphi = \beta_{P-F} V^{1/2} \qquad (3.14)$$

Here V is an applied voltage, β_{P-F} is the Poole–Frenkel coefficient:

$$\beta_{P-F} = \left(\frac{q^3}{\pi\varepsilon\varepsilon_0 d}\right)^{1/2} \qquad (3.15)$$

As can be seen from equation (3.15), *I–V* characteristics should be linear in the $\ln(J) - \sqrt{V}$ coordinates. Indeed, the reverse *I–V* characteristics of p-(001)Si–LiNbO₃–Al heterostructures are linear in $\ln(J) - \sqrt{V}$ coordinates at relatively high voltages, as shown in figure 3.8. As regards the low voltage nonlinear section, apparently it is associated with contact-limited conductivity through the Si/LiNbO$_3$ interface.

The experimental Poole–Frenkel coefficient β_{P-F}^{ex} is derived from the slope of this linear section and it is very close to the theoretical coefficient β_{P-F}^{theor}, calculated using equation (3.15) (see table 3.1). Good agreement between the two coefficients β_{P-F}^{ex} and β_{P-F}^{theor} is evidence that the Poole–Frenkel emission is the prevalent charge transport mechanism at reverse biases in the studied heterostructures.

Thus, our preliminary study has demonstrated that LiNbO$_3$ films, fabricated by the RFMS method on p-(001)Si substrates, contain a relatively high concentration of CLC in their band gap. Significant leakage currents and the presence of positive oxide charge restrict the use of the C–V method and impedance spectroscopy because they are in the deep depletion regime close to the inversion regime, which can be a serious limitation to their practical application.

Taking into account these results, we will conduct all further research based on the heterostructures fabricated on n-Si substrates.

Figure 3.9 shows typical C–V characteristics of n-(001)Si–LiNbO$_3$–Al heterostructures fabricated by RFMS on n-Si substrates, which are also similar to those for MIS structures.

As can be seen from figure 3.9 at zero bias the studied heterostructures are in the accumulation regime. Thus, at direct biases ('+' at a metal electrode), these heterostructures can be analysed as MIS structures, as was done for I–V analysis at the direct biases. Also, it is important to note, that C–V curves are shifted to the left along the voltage axis relative to an ideal case. This fact suggests the positive

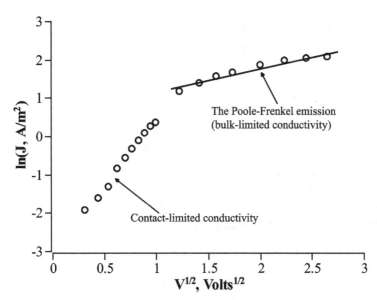

Figure 3.8. I–V characteristics of p-(001)Si–LiNbO$_3$–Al heterostructures in $\ln(J) - \sqrt{V}$ coordinates at reverse biases ('+' on Al).

oxide charge exists in LiNbO$_3$ film, which can be accounted for by the presence of a high defect concentration in the film. Using standard $C–V$ analysis [2] we have determined the dielectric constant $\varepsilon = 29$ and the effective density of positive charge $Q_{ef} = 3.8 \times 10^{-8}$ C cm^{-2} in the LiNbO$_3$ film. Besides the effective charge, sometimes it is important to determine the total (integral) charge in a dielectric Q_{ox} and its centroid position (center of mass position) d_c. For the studied heterostructures we have derived the following parameters: $Q_{ox} = 2.2 \times 10^{-6}$ C cm^{-2}, $d_c = 134$ nm.

Typical $I–V$ characteristics of the studied heterostructures, fabricated by the RFMS method on n-Si substrates are shown in figure 3.10 and can be described in the framework of MIS structures by the equation (3.3) similar to the analysis done for the heterostructures, formed on p-Si.

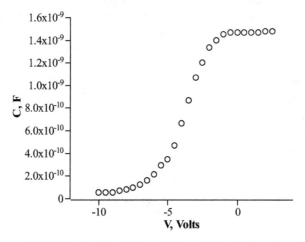

Figure 3.9. Typical high frequency (1 MHz) $C–V$ characteristics of (001)Si–LiNbO$_3$–Al heterostructures, fabricated by the RFMS method on n-Si substrate.

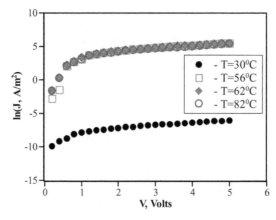

Figure 3.10. Typical $I–V$ characteristics of n-(001)Si–LiNbO$_3$–Al heterostructures, fabricated by RFMS at direct bias ('+' at Al contact) at different temperatures.

Following the procedure described above, we have estimated the concentration of CLC in the LiNbO$_3$ films through the ideality factor of the I–V curve at 30 °C using equation (3.7), obtaining the concentration of $N_t = 7 \times 10^{17}$ cm^{-3}. Furthermore, as shown in figure 3.10, the I–V characteristics of the studied heterostructures are temperature independent in the range of $T = 56$–90 °C. One of the possible conductivity mechanisms can be the non-activated hopping conductivity. In the framework of this mechanism the I–V characteristic is described as [14]:

$$J = J_0 \exp(-(E_0/E)^{1/4}) \tag{3.16}$$

Here E is the applied electric field strength, J_0 the field-independent constant and E_0 is the characteristic field, defined by [15]:

$$E_0 = \frac{\lambda}{D(E)a^4 q} \tag{3.17}$$

Here $D(E)$ is the energy density of localized states near the Femi level, d the film thickness, a is the localization length, λ is a dimensionless constant, which is usually equal to ~16 [14].

The energy dispersion within the Fermi level is described by the following formula:

$$\Delta E = \frac{3}{2\pi D(E)R^3} \tag{3.18}$$

The average hopping distance in the case of a non-activated mechanism is defined as [16]:

$$R = \frac{1}{(D(E) \cdot q \cdot E)^{1/4}} \tag{3.19}$$

Thus, the bulk concentration can be determined through the following expression:

$$N_t = D(E)\Delta E \tag{3.20}$$

Indeed, the I–V curves shown in figure 3.10 are linear in $\ln(J) - E^{-1/4}$ coordinates (figure 3.11), which is in good agreement with equation (3.16) and our previous work [17].

Using the parameter E_0, derived from the slope of a linear part of I–V characteristics in figure 3.11, and equations (3.17)–(3.20), we get the following parameters, characterizing the hopping conductivity over CLC: $R = 60$ Å, $D(E) = 1.6 \times 10^{21}$ eV^{-1} cm^{-3}, $N_t = 2.3 \times 10^{18}$ cm^{-3} [17]. It is important to emphasize, that as it was for heterostructures, fabricated on p-Si substrates, the studied heterostructures also manifested relatively high concentrations of CLC in LiNbO$_3$ films. Therefore, the formation of this type of CLC is influenced by sputtering conditions rather than substrate type. A detailed study on the relationship between sputtering conditions and the parameters of CLC will be presented in the following chapter.

3.1.2 Ferroelectric properties of LiNbO₃ thin films

As mentioned in chapter 1, ferroelectric properties of $LiNbO_3$ are the most important for applications in memory units. Ferroelectric materials, having the remnant polarization, manifest ferroelectric hysteresis independent of polarization on the applied electric field (P–E loop). A method of observation of this phenomenon was first proposed by Sawyer and Tower [18]. The Sawyer–Tower circuit is shown in figure 3.12.

Two capacitors, C_F and C_0, connected in series, represent a ferroelectric (in our case $LiNbO_3$) and a linear integrating capacitor, respectively. Normally, $C_0 \gg C_F$ and hence almost the whole applied voltage V_i is dropped across C_F. The applied voltage V_i is applied to the horizontal plates of an oscilloscope, so the horizontal axis on the $P(E)$ graph represents the electric field, applied to the ferroelectric film:

$$E(t) = \frac{V_i(t) - V_0(t)}{d} \approx \frac{V_i(t)}{d} \qquad (3.21)$$

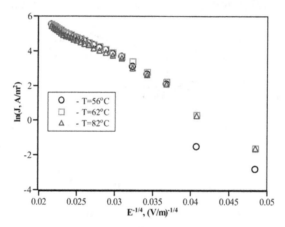

Figure 3.11. *I–V* characteristics of (001)Si–LiNbO₃ heterostructures fabricated by the RFMS method in coordinates $\ln(J) - E^{-1/4}$ at different temperatures.

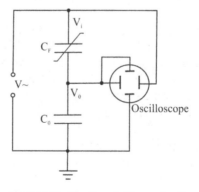

Figure 3.12. The Sawyer–Tower circuit.

Here d is the film thickness. The voltage applied to C_0 is recorded by the vertical plates of the oscilloscope and it is proportional to polarization of the ferroelectric capacitor C_F:

$$P(t) = \sigma(t) = \frac{C_0 V_0(t)}{S} \qquad (3.22)$$

$\sigma(t)$ is the surface charge density on C_F and S is the electrode area. Thus, the P–E loop is registered on the oscilloscope screen. In fact, equations (3.21) and (3.22) are applied only to the metal–ferroelectric–metal system. For the metal–ferroelectric–semiconductor system euqation (3.21) can be modified as:

$$E(t) = \frac{V_i(t) - \phi_s}{d} \qquad (3.23)$$

where ϕ_s is the surface potential at the ferroelectric–semiconductor interface. Moreover, because thin LiNbO$_3$ film is not an ideal insulator, current leakage can take place, leading to a change in $\sigma(t)$.

Figure 3.13 shows $P(E)$ characteristics of (001)Si–LiNbO$_3$ heterostructures fabricated by the RFMS method without the ion assist effect.

As can be seen from figure 3.13, all P–E loops demonstrate the saturation to the remnant polarization around $P_r = 14 \ \mu\text{C cm}^{-2}$. All loops are asymmetric (shifted to the right along the horizontal axis), and the coercive field depends on the amplitude of the applied electric field (this dependence is shown in figure 3.14). This shift can be caused by the presence of built-in fields E_b in the studied films which also depend on the applied field magnitude E_m.

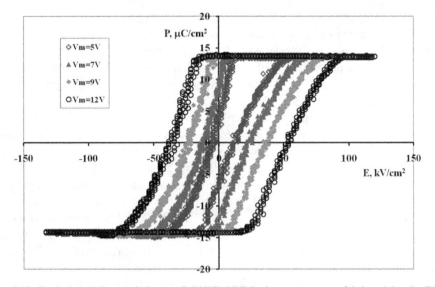

Figure 3.13. Typical P–E hysteresis loops of (001)Si–LiNbO$_3$ heterostructures, fabricated by the RFMS method without the ion assist effect and recorded at different amplitudes of the applied voltage V_m.

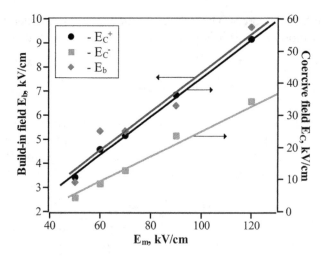

Figure 3.14. Dependence of the coercive field E_c and the built-in field E_b on the magnitude of the applied field E_m for LiNbO$_3$ films, fabricated by the FMS method without the ion assist effect. E_c^+ and E_c^- are the coercive fields, measured in positive and negative electric field range, respectively.

There are some works [19–21] which persuasively demonstrate that the shift of P–E curves is triggered by various factors, associated not only with properties of films but also with electric measurements.

According to the Sawyer–Tower method, any shift along the horizontal axis, representing an electric field in a sample, corresponds to the presence of built-in field (DC voltage) in the film which is added to the applied external electric field.

The built-in field can result from the fact that during the growing process of ferroelectric films the layers having different composition can be formed at the top and bottom surface of a film due to technological aspects (gradients of temperature and defect concentration), which can be the source of P–E loops shift [19, 20]. One reason for that can be the gradient of concentration of CLC in a film (oxygen vacancies, for example). To support this argument, some authors demonstrated that P–E hysteresis loop shift is, to a great extent, affected by inhomogeneity in the charge distribution in the film [22, 23].

Some investigators proposed an explanation of this shift based on the dipoles existing in a film. These dipoles are formed by the charged vacancies contributing to the depolarization process and blocking the switching in ferroelectric films [23]. However, other researchers argue, that the observed shift of P–E curves is not associated with dipole orientation, rather, it arises from trapping of electrons by vacancies [21, 24]. Distribution of the dipole ensembles changes under external electric field, modulating potential walls for the trapped electrons; in other words, polarization is a driving force of the charge capture. This approach is in agreement with figure 3.14, which demonstrates a strong linear dependence of the shift in P–E loops on the amplitude of applied field with slope, agreeing with [24].

Thus, we can reasonably assume that oxygen vacancies or other charged defects, existing at the interfaces, change their distribution during the growing process, forming space charge, influencing electrical properties and, specifically, polarization of LiNbO$_3$ films.

Earlier in this chapter, based on C–V analysis, we determined the total (integral) charge in a dielectric Q_{ox} and its centroid position, located at the distance d_c from the surface of LiNbO$_3$ film (see figure 3.15).

We need to briefly explain the classification of charge in a dielectric. According to [25] the oxide charge can be divided into four types: the fixed oxide charge (Q_f), the mobile ionic charge (Q_{mi}), the interface-trapped charge (Q_{it}), the oxide-trapped charge (Q_{ot}), and the main properties of these charges are given in table 3.2.

Taking into account the results of C–V analysis, (see discussion above and [26]), we can conclude, that the oxide charge Q_{ox} in our heterostructures is positive and does not depend on the substrate type. Evidently, Q_{ox} is not the interface-trapped charge (Q_{it}), the sign and value of which are strongly influenced by the type of substrate. At this stage we only can deduce that Q_{ox} in the studied LiNbO$_3$ films is a sum $Q_{ox} = Q_{ot} + Q_f + Q_{mi}$. The issue of distinguishing the contribution of each charge type is very complicated and needs to be investigated in a separate study. For example, thermal annealing of the studied heterostructures could help to clear up this issue.

Apparently, the trapped charge forms the space charge in LiNbO$_3$ film, causing the shift in P–E loops. A model developed in [27], introduces thin non-switchable layers called 'passive layers' which are formed near the electrodes and have properties that differ from those in the bulk of a material. The fact that these layers have finite conductivity and contain space charge can be crucial in switching of ferroelectric films. Obviously, the space charge asymmetrically trapped in LiNbO$_3$ films should lead to non-symmetric switching. Primarily, this charge is trapped by the states distributed near the electrode surface. In this case, the studied film can be modeled as an ideal ferroelectric capacitor and another capacitor (a 'surface

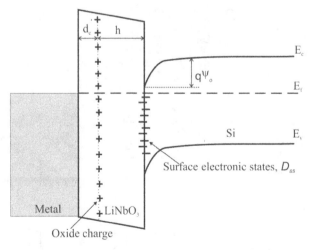

Figure 3.15. Schematic band diagram of metal–LiNbO$_3$–Si heterostructure with positive oxide charge, distributed at the distance of d_c from the film surface.

Table 3.2. Characteristic of different types of oxide charge.

Property	Types of charge			
	Interface-trapped charge (Q_{it})	Oxide-trapped charge (Q_{ot})	Fixed oxide charge (Q_f)	Mobile ionic charge (Q_{mi})
Location	At the semiconductor–oxide interface	Within the bulk of an oxide	Near the semiconductor–oxide interface	Within the bulk of an oxide
Charges	Positive/negative	Positive/negative	Positive	Positive
Sources	Structural defects, metal impurities	Ionizing radiation, injection	Structural defects	Ionic impurities (Na^+, K^+, Li^+ ...)
Dependence on applied voltage V_G	Depends on V_G	Does not depend on V_G	Does not depend on V_G	Does not depend on V_G
Charging state	Charged and discharged by applying V_G	Charged and discharged under specific conditions	Fixed	Fixed (immobilized) at temperatures below 390 °C

capacitor'), with a thickness equal to the distance from the electrode surface to the centroid of the trapped charge and connected in series (see figure 3.16). The charge of that capacitor is equal to the total charge, sitting between two capacitors.

The total voltage drop and the continuity equation for the normal component of the electric displacement D in this structure are described by the following equations:

$$V = E_f h + E_d(t)d_c$$
$$D_f - \sigma = \varepsilon_0 \varepsilon_d E_d(t) \tag{3.24}$$

Here d_c and h are the thickness of the passive layer and the ferroelectric layer, respectively. E_f and E_d are the electric fields in the ferroelectric capacitor and passive layer, ε_d is the dielectric constant of the passive layer and σ is the the surface trapped charge density at the interface between the passive and ferroelectric layers. Combining both equations we get:

$$D_f(t) - \sigma = \frac{\varepsilon_d \varepsilon_0 (V - E_f h)}{d_c} \tag{3.25}$$

In [27] the offset voltage (V_{off}) is the difference in applied voltage, which produces the same switching of a structure (i.e. the same values E_f and D_f are attained) at both charged and neutral 'passive layer/ferroelectric' interface. In this case it follows from equation (3.25) that

$$V_{\text{off}} = -\frac{\sigma d_c}{\varepsilon_d \varepsilon_0} \tag{3.26}$$

Thus, the offset voltage directly depends on σ and d_c. In [27], the following expression for the dielectric constant of the passive layer was proposed:

$$\varepsilon_d = -\frac{\varepsilon_f d_{tot} \varepsilon_e}{h \varepsilon_e - \varepsilon_f d_{tot} - 2\varepsilon_f h} \tag{3.27}$$

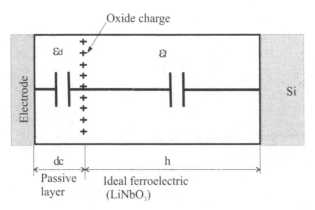

Figure 3.16. Schematic pattern of an equivalent circuit of a ferroelectric film with the passive layer, triggered by positive charge, existing at a depth of d_c.

Here ε_e and d_{tot} are the effective dielectric constant and total thickness of a film ($d_{tot} = d_c + h$) and ε_f is the dielectric constant of a ferroelectric layer. Analysing $C–V$ characteristics we derive the following parameter values required for estimation of σ: $\varepsilon_f = 28$, $\varepsilon_e = 61$, $d_c = 134$ nm. Taking into account that for the studied film $d_{tot} = 1$ μm and the maximum of applied voltage has a magnitude of 12 V, we estimated the trapped charge density in the studied film through equations (3.26) and (3.27) as $\sigma = 4.4 \times 10^{-6}$ C cm^{-2}. This value is in good agreement with the oxide charge determined earlier from $C–V$ analysis ($Q_{ox} = 2.2 \times 10^{-6}$ C cm^{-2}).

Therefore, the most probable source of built-in field in the studied LiNbO$_3$ films, fabricated by RFMS and causing the $P–E$ loops to shift, is the oxide-trapped charge with density of $\sigma = 4.4 \times 10^{-6}$ C cm^{-2}.

$P–E$ hysteresis loops for LiNbO$_3$ films, fabricated by the IBS method, are analysed in our work [28] and they were analogous to those recorded for the films and deposited by RFMS (see figure 3.17).

$P–E$ loops parameters for the films, deposited by both methods are listed in table 3.3.

It is worth noting, that the remnant polarization P_r, derived from $P–E$ loops (see table 3.3) is considerably less than those for bulk lithium niobate ($P_r = 71$ μC cm^{-2}), and close to the values, reported in [3, 29]. It can be concluded from the fact that the studied films, deposited without the ion assist effect, had an arbitrary orientation of polycrystalline grains (see chapter 2) and consequently demonstrated the lower ability of domains to be oriented in external electric field. We will recall this issue in chapter 4 in terms of the sputtering condition effect on ferroelectric properties of LiNbO$_3$ films. Regarding the films, deposited by IBS method, we have revealed the Fermi level pinning due to high density of surface states at the Si/LiNbO$_3$ interface, which does not allow the correct estimation of Q_{ef} in these films [12].

To sum up, based on the above results, we determined the main electrical parameters of LiNbO$_3$ films, fabricated by IBS and RFMS methods *without the ion assist effect*, which are presented in table 3.4.

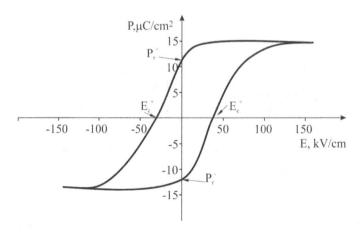

Figure 3.17. Typical hysteresis $P–E$ loops of LiNbO$_3$ films, fabricated by the IBS method.

Table 3.3. Parameters of experimental P–E loops for LiNbO$_3$ films, deposited by RFMS and IBS methods onto (001)Si substrates.

Deposition method	Degree of crystallinity	P_r^+ ($\mu C\ cm^{-2}$)	P_r^- ($\mu C\ cm^{-2}$)	E_c^+ ($kV\ cm^{-1}$)	E_c^- ($kV\ cm^{-1}$)	Built-in field, E_b ($kV\ cm^{-1}$)
IBS	Polycrystalline	11.2	−12.4	+29	−29	0
RFMS (without the ion assist)	with the random grain orientation	13.7	−14.2	+53.6	−34.3	9.7

Table 3.4. Basic electrical properties of LiNbO$_3$ films in Si–LiNbO$_3$ heterostructures.

Substrate	p-Si	n-Si	n-Si
Deposition technique	RFMS (without the ion assist)		IBS
Dielectric constant ε	28	29	29
Resistivity, ρ (Ohm cm)	1×10^9	1×10^9	3×10^9
Remnant polarization, P_r ($\mu C\ cm^{-2}$)	–	14.0	11.7
Coercive field, E_c ($kV\ cm^{-1}$)	–	44.0	29.0
Effective charge in LiNbO$_3$ film, Q_{ef} ($C\ cm^{-2}$)	3.8×10^{-8}	3.8×10^{-8}	–
Concentration of CLC, N_t (cm^{-3})	2.5×10^{17}	7.1×10^{17}	–
Energy of CLC E_t (eV)	0.27	–	–

3.2 Conduction mechanisms in Si–LiNbO$_3$ heterostructures

As pointed out in chapter 1, the electrical properties of LiNbO$_3$-based heterostructures, being the essential parts of integrated electronics devices, are affected by charge transport in these heterostructures.

It is important to note that many applications of thin films are based on the application of an external electric field to a ferroelectric capacitor, causing leakage currents. If these currents are significant, they do not allow the polarization switching to be observed. Solutions, reducing these leakage currents, can be found only through a deep understanding of conductivity mechanisms and the influence of a current on the parameters of films and heterostructures. Specifically, leakage currents can influence the shape of the P–E loop, since this curve is derived through integration of charge, released in the polarization reversal process. High leakage currents suppress the hysteresis, masking ferroelectric properties in the studied sample. Conductivity mechanisms are affected by microstructures, defects, composition uniformity, external fields, temperature and charge states at the interfaces. The majority of perovskite ferroelectrics can be considered as semiconductors with low carrier mobility [30]. When a semiconductor is connected to a metal (for example in the formation of metallic contacts), the Schottky barrier is formed with a space-charge region or with a depletion region, leading to band bending at the interface.

Experimental results, demonstrating decrease in coercive field with film thickness, agree with the models of domain walls pinning and their formation in the depletion regions [31]. Some authors argue [32, 33] that oxygen vacancies, accumulating at the interface electrode/film, play a crucial role in the polarization reversal process. Specifically, regions with oxygen reduction can grow towards the bulk, effectively shielding a film from the applied voltage, causing loss of polarization. Taking into account these facts, there is no doubt that conduction mechanisms in (001)Si–LiNbO$_3$ heterostructures is a powerful tool for detailed study of their electrical properties.

For detailed investigation of conduction mechanisms in (001)Si–LiNbO$_3$ hetero-structures, single phase LiNbO$_3$ films were deposited by RFMS and IBS methods onto n-Si substrates according to the optimal sputtering regimes, developed in chapter 2. Below we present the analysis of DC conductivity in heterostructures, fabricated by the IBS method, whereas conductivity of heterostructures, synthesized by the RFMS technique, will be discussed in the section on their band diagram.

According to the results of C–V analysis, reported in the previous sections of this chapter, the studied (001)Si–LiNbO$_3$–Al heterostructures are under an accumulation regime at the positive biases ('+' at Al). This means, that a sufficient number of electrons are supplied to LiNbO$_3$ layer from a silicon substrate. Taking into account this fact, and that resistivity of the substrate, in our case, is considerably less than those for LiNbO$_3$ ($\rho = 3 \times 10^9$ Ohm cm—see table 3.4 and [26]), the studied heterostructures can be considered as a metal–insulator–metal system.

I–V characteristics of the studied heterostructures can be described by the special case of equation (3.3) [7]:

$$J = J_s\left(\exp\left(\frac{qV}{n \cdot kT}\right) - 1\right) \qquad (3.28)$$

Here J is the current density, q is the the electron charge, V the applied voltage, k is Boltzmann's constant, T temperature, n the ideality factor, defined by equation (3.6), and J_s is the saturation current density.

I–V characteristics were studied at the temperature range of 90–300 K [34] and shown in figure 3.18.

Two sections can be seen on the I–V curves (see figure 3.18): the first section (which we will call the region of low and average voltage ($0 < V < 1.5$ V) or in terms of electric fields $0 < E < 30$ kV cm^{-1}) apparently results from the contact phenomena and corresponds to the fast growing current, the second section, associated with high voltage ($V > 1.5$ V or $E > 30$ kV cm^{-1}), and is influenced by the bulk of the film.

The temperature dependence of current in the Arrhenius coordinates for voltages, corresponding to different sections of I–V curves are shown in figure 3.19. The activation energies of conductivity E_a, derived from the slopes of these graphs are also indicated in figure 3.19.

In the range of low voltages ($0 < V < 0.1$ V) BAX I–V characteristics are linear ($J \propto V$), which is evidence of the Ohmic conductivity (see inset in figure 3.18). The studied LiNbO$_3$ films are polycrystalline, containing, as a rule, a high concentration

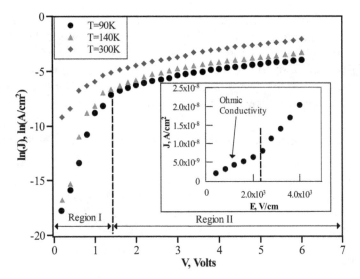

Figure 3.18. Typical *I–V* characteristics of the (001)Si–LiNbO₃ heterostructures at various temperatures.

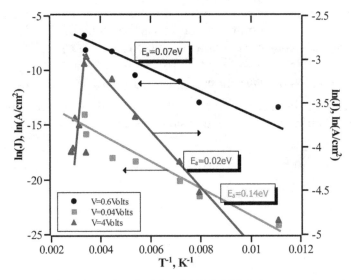

Figure 3.19. Temperature dependence of conductivity of (001)Si–LiNbO₃ heterostructures at low ($V = 0.04$ V), average ($V = 0.6$ V), and high ($V = 4$ V) voltage.

of CLC (traps) in the band gap. It this case the Ohmic conductivity can be realized by the hopping of electrons over CLC which agrees with results reported in [35] for hopping conductivity. *I–V* characteristics in the framework of this mechanism are described by the expression [7]:

$$J = \sigma_0 \frac{V}{d} \exp\left(-\frac{E_a}{kT}\right) \qquad (3.29)$$

Here σ_0 is a constant, E_a is the activation energy of conductivity, which is derived from the slope of the graph $\ln(J) - 1/T$ (see figure 3.19) and is equal to $E_a = 0.14$ eV, close to the value obtained in paper [35]. The activation energy of $E_a = 0.07$ eV (see figure 3.19) is also close to the value determined in the framework of electronic hopping conductivity [36]. For detailed analysis of conduction mechanisms let us consider each voltage and temperature range on the I–V curve.

Region of average electric fields in the temperature interval of $T = 90$–140 K

Extremely weak temperature dependence of conductivity in the range of average electric fields ($3 < E < 30$ kV cm^{-1}) at temperatures of $T = 90$–140 K (see figure 3.19) can be attributed to Fowler–Nordheim tunneling. In this case current density obeys equation (1.10) (see chapter 1). According to equation (1.10) I–V characteristic should be linear in the coordinates $\ln(J/E^2) - 1/E$ (the Fowler–Nordheim coordinates) and from the slope of this linear dependence the potential barrier height ϕ can be derived through equation (1.11) (see figure 3.20).

Because the tunneling through the thick films is improbable, electrons are able to tunnel through the potential barrier between the metal electrode and the nearest trap centers, as was demonstrated in chapter 1 (see figure 1.17). Thus, if ϕ is the average potential barrier height, the energy position of traps in the band gap E_g of LiNbO$_3$ with respect to the conduction band bottom can be found as $E_t = E_g/2 - 2\phi$. In our particular case we obtained the following value: $E_t = 1.76$ eV.

Region of average electric fields in the temperature range of $T = 140$–300 K

I–V characteristics in this interval can be described in the framework of the Richardson–Schottky emission by the Simmons formula (1.8) (see chapter 1). In

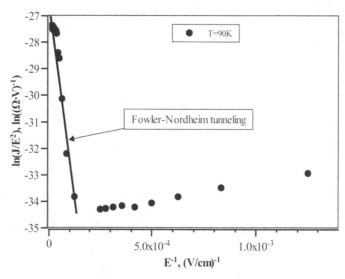

Figure 3.20. I–V characteristics of the (001)Si–LiNbO$_3$ heterostructures in the Fowler–Nordheim coordinates. The rectilinear site corresponds to the Fowler–Nordheim tunneling.

this case, I–V characteristics should be linear in the Simmons coordinates $(\ln(J/(ET^{3/2}))$ versus $E^{1/2})$ with the slope of β/kT (β—coefficient, denoted by equation (1.9) from chapter 1). Moreover, the slope of temperature dependence of the Richardson constant in the Arrhenius coordinates ($\ln(A^*)$ versus $1/T$) gives the potential barrier height φ_0. In fact, I–V characteristics of the studied heterostructures in the range of fields $2 < E < 30 \text{ kV cm}^{-1}$ are linear in the Simmons coordinates, as shown in figure 3.21.

Analysing these graphs yields the barrier height value of $\varphi_0 = 0.26$ eV.

Region of high electric fields

As was noted above, in the range of high electric fields ($E > 30 \text{ kV cm}^{-1}$) conductivity of the studied heterostructures is influenced by bulk properties of LiNbO$_3$ films. The conductivity of polycrystalline LiNbO$_3$ films is affected by the properties of grain boundaries rather than their volume [17]. Formation of potential barriers at the interfaces is one of the key factors influencing electrical properties of polycrystalline films and it is usually modeled by a double barrier, similar to the Schottky barrier (figure 3.22). Such heterostructures are complex in terms of description due to the variety of conduction mechanisms, some of which are shown in figure 3.22.

In the framework of the thermally-assisted tunneling mechanism through the potential barrier at the grain boundaries, the value of J_s and the factor n in equation (3.28) are given by [37]:

$$J_s = \frac{A \cdot T}{k} \sqrt{E_{oo}\pi} \sqrt{\frac{q\varphi_b}{\cosh\left((E_{oo}/kT)^2\right)}} \exp\left(-\frac{q\varphi_b}{E_{oo} \coth\left(E_{oo}/kT\right)}\right) \qquad (3.30)$$

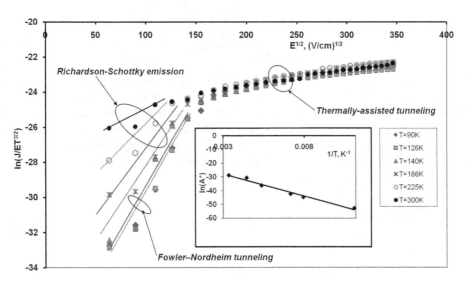

Figure 3.21. I–V characteristics of the (001)Si–LiNbO$_3$ heterostructures in the Simmons coordinates at various temperatures. The temperature dependence of the pre-exponential factor in expression (1.8) is given in the inset.

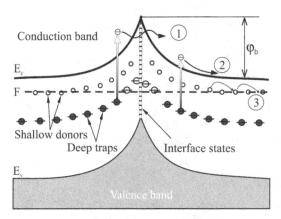

Figure 3.22. Energy diagram of the grains' interface in a polycrystal and some possible charge transport mechanisms. Here, φ_b is the height of the potential barrier, F is the Fermi level, 1 is the thermally assisted tunneling through a potential barrier, 2 is the carrier emission from the deep levels, and 3 is the hopping conduction mechanism.

$$n = \frac{T_c}{T}\left(1 + \frac{d}{L}\right) \qquad (3.31)$$

Here A is the Richardson constant, φ_b and L are the effective potential barrier height and the width of a depletion area at the grain boundaries, d is the dielectric layer thickness. Parameters E_{oo} and T_c are defined by the formulas:

$$E_{oo} = \frac{qh}{4\pi}\sqrt{\frac{N_d}{m^*\varepsilon\varepsilon_0}} \qquad (3.32)$$

$$T_c = \frac{E_{oo}}{k}\coth\left(\frac{E_{oo}}{kT}\right) \qquad (3.33)$$

Here m^* the carrier effective mass, ε and ε_0 the dielectric permittivity of a material and the electric constant, N_d is the concentration of ionized donors in the layer. Parameter E_{oo} can be used to estimate the contribution of the thermally assisted tunneling in conductivity. Specifically, if $E_{oo}/kT \sim 1$ the tunneling component prevails, otherwise if $E_{oo}/kT \ll 1$, the thermionic emission dominates. The width of a potential barrier and concentration of charge states N_{is} at the grain boundaries can be estimated through the expressions [37]:

$$\varphi_b = \frac{qN_dL^2}{2\varepsilon\varepsilon_0}\left(1 - \frac{2L}{3d_g}\right) \qquad (3.34)$$

$$N_{is} = \sqrt{\frac{2N_d\varepsilon\varepsilon_0\varphi_b}{q}} \qquad (3.35)$$

Here d_g is the average grain size. Determining J_s and n from experimental I–V characteristics we have solved the system of equations (3.30), (3.31), (3.34) for φ_b, L and N_d and the results are presented in table 3.5.

Bulk concentration of traps recalculated from N_{is} is relatively high with a value of $N_t = 9.0 \times 10^{19}$ cm^{-3}. As seen from figure 3.23, there is good agreement between the experimental I–V curve and the theoretical one, calculated from equations (1.11), (3.28), (3.30)–(3.34) [34], and also the temperature dependence of a pre-exponential factor J_s in equation (3.30), which is shown in the inset of figure 3.23.

The section with negative temperature coefficient in the Arrhenius graph (see the inset in figure 3.19) is of interest. It can be explained by the fact that the grain boundaries affect conductivity of LiNbO$_3$ films at strong electric fields. The charge states, accumulated at the grain boundaries in LiNbO$_3$ films, trap electrons from the bulk of grains, leading to the formation of depletion areas near the grain boundaries and barriers, controlled by electronic traps. Electrons, overcoming the barriers,

Table 3.5. Results of I–V analysis of (001)Si–LiNbO$_3$ heterostructures, fabricated by IBS method.

Average grain size d_g, nm	Depletion zone's width at the intergranular interface, L (nm)	Potential barrier's effective height, φ_b (eV)	Concentration of ionized donors, N_d (cm^{-3})	Density of states, N_{is} (cm^{-2})
50	9	0.7	3×10^{19}	2.5×10^{13}

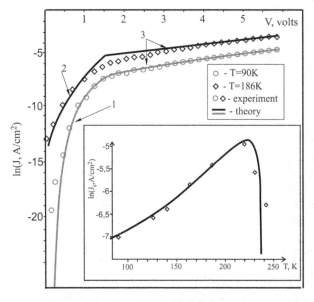

Figure 3.23. The typical I–V characteristics of the (001)Si–LiNbO$_3$ heterostructures at the temperature of 90 K and 186 K. Data labeled as 1 is calculated from the equation (1.10) (see chapter 1) at $T = 90$ K, curve 2 results from the equation (1.8) (see chapter 1) at $T = 186$ K, 3 is calculated from (3.28) and (3.30)–(3.31). The inset shows the temperature dependence on the pre-exponential factor in equation (3.30).

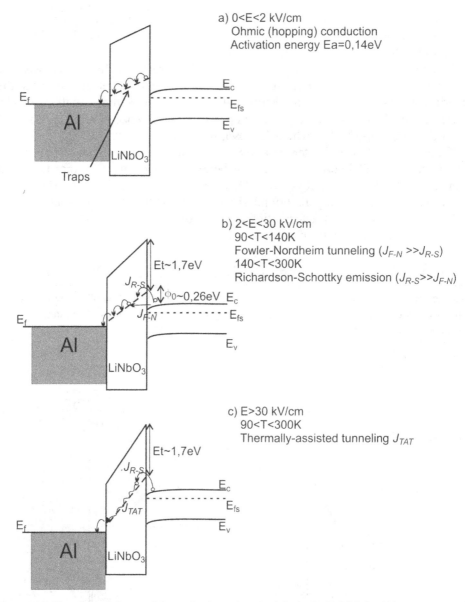

Figure 3.24. The schematic image of the mechanisms of conductivity in the Si–LiNbO₃–Al heterostructures in various intervals of voltage and temperatures [34]. Reprinted by permission from Springer Nature. Copyright 2013.

penetrate in the intergranular layers and fill traps first and then form the space charges within grain boundary areas. These space charges can overcome the potential barrier at the interface and move out from the states to the conduction zone, creating the current. This electronic current rises with temperature and with increase of kinetic energy of electrons. Therefore, the number of space charges in the inter-grain areas decline and the electronic states are filled by carriers from deeper trap centers. It determines the nature of the near-grain layers which those carriers should overcome to maintain conductivity. Therefore, the barrier height, overpassed by charge carriers, increases with temperature, leading to the increase in current.

Summarizing the analysis of I–V characteristics presented above, the conduction mechanisms in the studied heterostructures, fabricated by IBS method, are shown schematically in figure 3.24 [34].

3.3 Band diagram of the Si–LiNbO₃ heterostructures

It is generally accepted that the design and analysis of a band diagram is a powerful tool for describing the engineering of the electrical properties of synthesized semiconductor heterostructures. One of the most popular approaches in construction of the band diagram is the electron affinity model [38]. Electron affinity is defined as the energy required to replace an electron from the bottom of the conduction band of an uncharged crystal to infinity. In the framework of this model, when a heterojunction is created, the conduction band offset is the difference between the electron affinities of two materials in the heterostructure:

$$\Delta E_c = \chi_1 - \chi_2 \qquad (3.36)$$

Taking into account that $\chi_{Si} = 4.05$ eV and $\chi_{LiNbO3} = 1.5$ [39], the conduction band offset is equal to $\Delta E_c = 2.55$ eV. However, this approach leads to an inappropriate result for ΔE_c contradicting the experimental data, especially for two materials with considerably different band gaps. The reason for this is that the electron affinity depends on surface charge and dipoles that have to be taken into account when ΔE_c is calculated. Moreover, when two materials are brought into contact, they exchange the charge via surface states. Thus, other models allowing estimation of the conduction band offset were proposed. An interesting and efficient model, based on the theory of quantum dipoles was proposed by Tersoff [40]. This model declares the existence of 'the interface induced gap states' that are similar to those, formed at the metal–semiconductor interface. From this point of view some heterostructures can be represented by two Schottky barriers connected in series, and thus, the conduction band offset can be determined as the difference in the Schottky barrier heights. As a result, the real band diagram of a heterostructure is influenced by a wide range of factors such as built-in charge, interface states etc, that have not been considered in the traditional Anderson model. As was emphasized by some authors, when surface states are presumably present at the metal–semiconductor heterojunction, the barrier height φ_b increases proportionally to the work function of a metal ϕ_m. It was proposed [41] that the slope $S = d\varphi_b/d\varphi_m$ is affected only by the density of states and can be approximated by the following expression:

$$(1/S - 1) = 0.1(\varepsilon - 1)^2 \qquad (3.37)$$

Here ε is the dielectric constant of a material. Consequently, there are two special cases: when $S = 1$ the Schottky barrier takes place, otherwise, when $S = 0$ the Bardin barrier is formed, which is totally independent of the metal used as an electrode.

Implying $\varepsilon = 28$ [26], the slope S was estimated to be 0.01, suggesting that in the case of Si–LiNbO$_3$ heterostructures the Tersoff model is more realistic. One possible band diagram of Si–LiNbO$_3$ heterostructures proposed in our recent work, is shown in figure 3.25 [42].

In the following section we will determine the parameters of the band diagram shown in figure 3.25 using I–V and C–V methods in the temperature range 80–300 K. We have deposited films with thickness of 0.5 μm by RFMS method onto n-Si substrates ($\rho = 4.5$ Ohm cm) under the optimal regimes with the ion assist, developed in chapter 2, ensuring the formation of the single-phase LiNbO$_3$ c-oriented films (see chapter 2, table 2.6) with a growth rate of 10 nm min^{-1}. Indeed, it follows from XRD patterns, that single-phase <0001> textured polycrystalline LiNbO$_3$ films are formed under these conditions due to the ion-assist effect (figure 3.26).

For successful building up of the band diagram, first information regarding the band gaps of materials in the heterostructure is needed. To determine the band gap of our LiNbO$_3$ films, we investigated the optical absorption in the region of the fundamental band edge. For this purpose we deposited LiNbO$_3$ films under the same regimes onto fluorphlogopite substrates ensuring transparency of the studied heterostructures in the visible wavelength range. The absorption spectra (the absorption coefficient α versus the incident photon energy) is shown in figure 3.27.

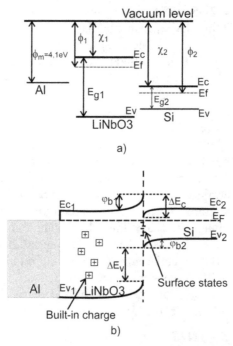

Figure 3.25. Band diagram of the (001)Si–LiNbO$_3$–Al heterostructure: (a) separated materials, (b) materials in a contact. Reprinted from [42] with permission from Elsevier. Copyright 2013.

It can be seen from figure 3.27 that the absorption coefficient rises steadily with photon energy and then increases sharply in the energy range 4.2–4.6 eV. Depending on the band structure of a semiconductor, the absorption coefficient obeys various laws and the frequency dependence $\alpha(\nu)$ can be expressed as [43]:

$$\alpha(\nu) \propto B\left(h\nu - E_g\right)^r \tag{3.38}$$

Here B is a constant, h is Planck's constant, E_g is the band gap, $r = 1/2$ for direct optical transitions, $r = 2$ for indirect ones. It was demonstrated [44, 45] that the band gap of single crystals of $LiNbO_3$ corresponds to both the direct and indirect optical transitions. Thus, based on equation (3.38) we can determine the direct band energy E_g^{dir} as an intercept of the linear section of the graph $\alpha^2(h\nu)$ with the horizontal axis

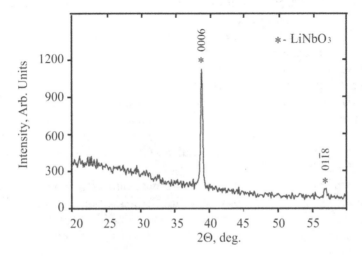

Figure 3.26. X-ray diffraction patterns of the $LiNbO_3$ films 0.5 μm thick formed on the (001)Si substrate.

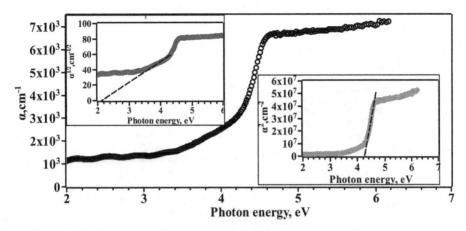

Figure 3.27. Dependence of the absorption coefficient α with the incident photon energy for the $LiNbO_3$ films. Insets show the dependence of α^2 and $\alpha^{1/2}$ with incident photon energy, respectively.

(see inset in figure 3.27). For the direct band energy we obtained the following result $E_g^{dir} = 4.2$ eV which is close to one for the bulk lithium niobate ($E_g^{dir} = 4.12$ eV [45]). The linear part, the steeper slope in figure 3.27, is attributed to the broad absorption band. The graph $\alpha^{1/2}(h\nu)$ (see inset in figure 3.27) is a linear function in the range of incident photon energies of 3.5–4.2 eV, which is evidence for indirect optical transitions [43]. The intercept of this linear graph with horizontal axis corresponds to the edge of the indirect band gap having, in our case, a value of $E_g^{ind} = 2.2$ eV. In the theory of small polarons, the absorption band edge is connected to the activation energy of conductivity W as $E_{opt} = 4\,W$ [46]. Accepting that in our case $E_{opt} = 2.2$ eV we estimate the activation energy as $W \approx 0.5$ eV which is slightly higher than that obtained in our previous work from ac conductivity of Si–LiNbO$_3$ heterostructures in the framework of variable range hopping conductivity [47]. To determine the band bendings, the Fermi level position and concentration of carriers, we applied methods of I–V and high-frequency C–V analysis.

Figure 3.28 shows the typical high-frequency ($f = 10^5$ Hz) C–V characteristic of a (001)Si–LiNbO$_3$ heterostructure. Similar to the heterostructures, analysed above, the C–V curve is shifted to the left along the voltage axis, that indicates the presence of a positive oxide charge in LiNbO$_3$ which is in total agreement with our previous results and the results of other investigators [48]. Also, as can be seen in figure 3.28 the studied heterostructures are under the accumulation regime at zero bias. According to the standard methodology [2] we re-drew the C–V characteristics in $(S/C)^2$ versus V coordinates (S is the metal contact area). This dependence is shown in figure 3.28 (see inset).

Since the graph corresponding to the uniform doping distribution is nonlinear, we have non-uniform distribution of donors in the Si substrate in the studied heterostructures. The concentration distribution is determined by graphical differentiation of the experimental $(S/C)^2 - V$ curve using the following well known expression [2]:

$$N_d(x) = -\frac{2}{q\varepsilon\varepsilon_0 S^2}\left(\frac{d}{dV}\left(\frac{S}{C(V)}\right)^2\right)^{-1} \tag{3.39}$$

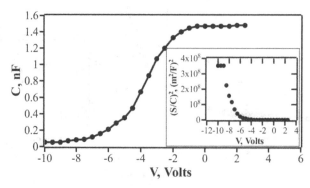

Figure 3.28. High-frequency ($f = 1 \times 10^5$ Hz) C–V characteristics of the (001)Si–LiNbO$_3$ heterostructures. The inset displays the dependence $(S/C)^2$ versus V for the studied heterostructures.

Here ε_0 is the electric constant, ε is the dielectric constant of a semiconductor, S is the area of a metal contact. Our calculations suggest that the concentration of ionized donors decreases from 5×10^{17} cm^{-3} at the Si/LiNbO$_3$ interface to 1×10^{15} cm^{-3} at a distance of 1 μm, that corresponds to the nominal donor concentration in the substrate (figure 3.29).

One of the origins of such a distribution can be the diffusion of Li atoms onto the substrate in the process of deposition of LiNbO$_3$ films. Lithium, having relatively high diffusion coefficient ($D = 2 \times 10^{-11}$ cm^2 s^{-1}), can penetrate deeply into Si, generating shallow donors [49, 50]. This is indicative, shallow donors are not observed in a silicon substrate in Si–SiO$_2$–LiNbO$_3$ heterostructures. Apparently, it is caused by the presence of the SiO$_2$ layer, preventing lithium diffusion in the substrate in the process of RFMS. Moreover, as was mentioned above, surface states influence heterojunction formation. Following the methods of C–V analysis [2] we obtain the energy distribution of surface states $D_{ss}(E)$ at the Si/LiNbO$_3$ interface, which is presented in figure 3.30.

The integral charge in the LiNbO$_3$ layer, being the sum of the total oxide charge Q_{ox} and the surface states charge with density of D_{ss} is determined using the following expression [2]:

$$Q_{sc}(\psi_s = \psi_{s0}) - Q_{ox}(\psi_s = 0) = q \int_0^{\psi_{s0}} D_{ss}(\psi) \cdot f_0(\psi) \; \mathrm{d}\psi \qquad (3.40)$$

Here ψ_{s0} is the surface potential at the regime of flat bands, Q_{sc} is the charge of the depletion zone in Si, D_{ss} is the density of surface states and $f_0(\psi)$ is the distribution function. All the results obtained from C–V analysis are given in table 3.6.

Figure 3.31 shows the I–V characteristics of the studied heterostructures at different temperatures. Two regions are clearly seen in I–V curves: region I of moderate electric fields (5–30 kV cm^{-1}) and region II of high electric fields (30–90 kV cm^{-1}).

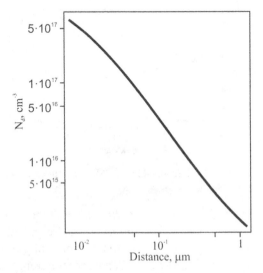

Figure 3.29. Doping profile of the (001)Si–LiNbO$_3$ heterostructure; x is the distance from Si/ LiNbO$_3$ interface towards the bulk.

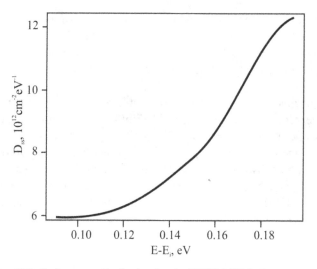

Figure 3.30. Surface-state distribution for the (001)Si–LiNbO$_3$ heterostructure.

Table 3.6. C–V analysis results for the (001)Si–LiNbO$_3$ heterostructures.

Donor concentration in Si substrate, N_d (cm^{-3})	1×10^{15}
Flat band voltage, V_{fb} (V)	-5.6
Fermi level position in Si substrate, $E_C - E_F$ (eV)	0.31
Surface band bending, ψ_{s0} (eV)	0.36
Effective charge in LiNbO$_3$ layer, Q_{eff} (C cm^{-2})	$+7.8 \times 10^{-7}$
Built-in charge in LiNbO$_3$ layer, Q_{ox} (C cm^{-2})	$+2.2 \times 10^{-6}$

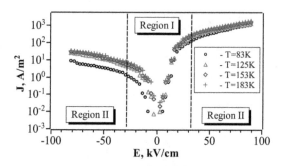

Figure 3.31. I–V characteristics of the Si(001)–LiNbO$_3$–Al heterostructures at different temperatures.

It is commonly stressed, that at a particular voltage I–V characteristics of heterostructures are influenced by the contact-limited conduction (see chapter 1). Recall that when conductivity depends on the carrier mobility in a semiconductor layer, in the framework of the Richardson–Schottky emission, the I–V curves should be linear in the coordinates $\ln(J/(ET^{3/2})) - E^{1/2}$ (Simmons coordinates) and can be described by the Simmons–Schottky equation (1.8) of chapter 1, which we give again here:

$$J_{R-S} = 2q\left(\frac{2\pi m^* kT}{h^2}\right)^{3/2} \mu E \exp\left(-\frac{q\varphi}{kT}\right)\exp\left(\frac{\beta\sqrt{E}}{kT}\right) \qquad (3.41)$$

On the other hand, when only over-barrier emission dominates, the Richardson–Schottky formula takes place [51]:

$$J = \frac{4\pi qm}{h^3}(kT)^2 \exp\left(-\frac{q\varphi}{kT}\right)\exp\left(\frac{\beta\sqrt{E}}{kT}\right) \qquad (3.42)$$

In equations (3.41) and (3.42) φ is the potential barrier height at the heterojunction, E is the applied electric field, m^* and μ are the carrier effective mass and mobility, β is the coefficient, defined by equation (1.9). Equations (3.41) and (3.42) differ only by the pre-exponential factor, so the mobility and applied field are the critical parameters, defining which formula is applicable in each particular case. In our experiments I–V characteristics are linear at high direct biases (region II in figure 3.31) and they are linearized in the Simmons coordinates (see figure 3.32).

On the other hand, it follows from formula (3.41), that the graph $\ln(J/(ET^{3/2}))$ – $1/T$ should be a straight line with a slope of γ:

$$\gamma = \frac{-q\varphi + \beta\sqrt{E}}{k} \qquad (3.43)$$

As follows from equation (3.43), the barrier height φ and the factor β, directly connected with dielectric constant of a material, can be derived from the slope and intercept of a graph γ versus $E^{1/2}$ (see figure 3.33).

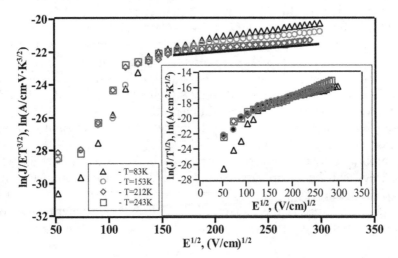

Figure 3.32. I–V characteristics of the (001)Si–LiNbO$_3$–Al heterostructures in the Simmons coordinates at direct bias ('+' on Al electrode). The inset shows the I–V characteristics in the Schottky coordinates at reverse bias ('−' on Al electrode).

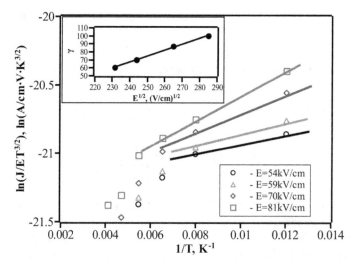

Figure 3.33. Temperature dependence of the (001)Si–LiNbO$_3$–Al heterostructure's conductivity in the Arrhenius coordinates in the framework of the Simmons–Schottky mechanism. The inset illustrates the field's dependence of the slope, related to the linear parts of graph $\ln(J/(ET^{3/2}))$ versus $1/T$.

As regards reverse biases ('–' at Al), I–V characteristics are linear in the Schottky coordinates (see inset in figure 3.32), indicating that equation (3.42) should be applied. To prove that the application of equations (3.41) and (3.42) is correct we have calculated the ratio of experimental to theoretical Schottky factors β_{ex}/β_{theor} for direct and reverse biases. Using formula (1.9) for β_{theor} and taking $\varepsilon = 28$, we obtained $\varphi = 0.02$ eV and $\beta_{ex}/\beta_{theor} = 1.03$ and $\beta_{ex}/\beta_{theor} = 1.09$ for direct and reverse biases, respectively.

In the range of moderate electric fields (region I in figure 3.31) the conductivity of the studied heterostructures is influenced by the Fowler–Nordheim tunneling through the barrier ϕ. In this case I–V characteristic should be linear in the Fowler–Nordheim coordinates $\ln(J/E^2)$ – $1/E$ (see chapter 1) and described by equation (1.10). The slope of this linear graph gives a barrier height of ϕ. In the range of moderate electric fields I–V curves are linear in the Fowler–Nordheim coordinates (figure 3.34) with a slope corresponding to the barrier height of $\phi = 0.03$ eV which is close to the height, obtained from I–V curves at high fields [42] and to the value, observed in the heterostructures, fabricated by the IBS method [34].

Figure 3.35 demonstrates good agreement between experimental data and I–V curves, calculated theoretically.

It is important to emphasize that potential barrier height, derived from I–V characteristics, can be an apparent one due to the significant influence of polarization of a ferroelectric, which can greatly reduce the barrier height in a heterojunction [52]. As a result, the apparent barrier height φ_{ap}, determined from I–V curves can be expressed through the actual value φ_b in the following way:

$$\varphi_{app} = \varphi_b - \sqrt{\frac{qP}{4\pi(\varepsilon\varepsilon_0)^2}} \qquad (3.44)$$

Here P is the ferroelectric polarization. Taking $P = 13.6$ µC cm^{-2}, which we obtained from the P–E hysteresis loop analysis [17] and using equation (3.44) we determine the actual potential barrier height $\varphi_b = 0.2$ eV.

According to the Tersoff model, the band offset in a heterojunction can be found as the difference between two Schottky barrier heights, so it is necessary to obtain information regarding the Fermi level position in the LiNbO$_3$ film of the studied heterostructures. It can be estimated using the following formula [2]:

$$E_F = \frac{kT}{q} \ln\left(\frac{N_c}{N_d}\right) \tag{3.45}$$

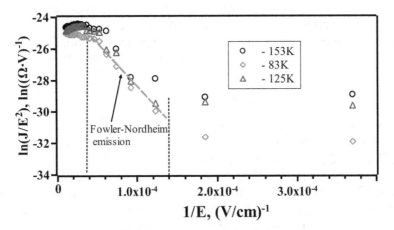

Figure 3.34. I–V characteristics of the Si(001)–LiNbO$_3$–Al heterostructures in the Fowler–Nordheim coordinates at the average electric field.

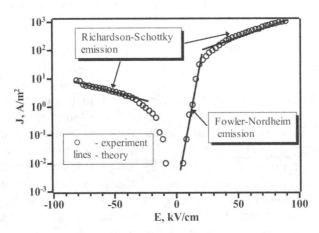

Figure 3.35. Experimental and theoretical I–V characteristics of the (001)Si–LiNbO$_3$–Al heterostructures at $T = 83$ K.

Table 3.7. Parameters of the band diagram of Si–LiNbO$_3$ heterostructures (see figure 3.25).

Band diagram parameter	Material in the heterojunction	
	LiNbO$_3$	Si
Band gap	$E_{g1} = 4.2$ eV	$E_{g2} = 1.1$ eV
Fermi level position	$E_{F1} = 0.33$ eV	$E_{F2} = 0.31$ eV
Band bending	$\varphi_{b1} = 0.2$ eV	$\varphi_{b2} = 0.36$ eV
Conduction band offset ΔE_C	0.6 eV	
Valence band offset ΔE_V	2.52 eV	

Here E_F is the position of the Fermi level relative to the conduction zone, N_d is the donor concentration, which in our case can be accepted as $N_d = 7 \times 10^{13}$ cm^{-3} [17] and N_c is the concentration of carriers in the conduction band. Thus, making use of equation (3.45) enables us to calculate the conduction band offset in the following way:

$$\Delta E_c = E_{F1} + \varphi_{b1} - (E_{F2} - \varphi_{b2}) \qquad (3.46)$$

In equation (3.46) we take into account the fact that according to C–V analysis at zero bias, the bands in Si bend down. All parameters of the proposed band diagram shown in figure 3.25 are presented in table 3.7.

It is important to note that the conduction and valence band offsets, listed in table 3.7, differ significantly from those calculated in the framework of the Anderson model through the electron affinities, indicating that interface states at the Si/LiNbO$_3$ heterojunction have a profound effect on formation of the studied heterostructure.

3.4 Impedance spectroscopy and ac conductivity of thin LiNbO$_3$ films

Impedance spectroscopy (IS) is one of the most informative tools for the investigation of the electrical properties of materials [53, 54]. IS is widely used for the study of dielectric materials (solid and liquid dielectrics whose electrical properties are caused by dipole orientation), materials with dominating electronic conductivity (crystalline and amorphous semiconductors, glasses, polymers), and also conducting dielectrics with ionic and electronic conductivity. Ac impedance spectroscopy is recorded in a wide frequency range, so different areas of a material are characterized according to their time constant (relaxation time). For polycrystalline materials, successfully distinguishing the response of the grain boundary from the bulk phenomena in a material depends on the appropriate equivalent circuit, describing the studied heterostructures. In general, the response of ac signal, applied on the studied sample, can be interpreted in the framework of one of the following formalisms: (1) the complex admittance $Y^* = 1/R_p + j\omega C_p$, (2) the complex impedance $Z^* = R_s + (j\omega C_s)^{-1}$, (3) the complex dielectric permittivity $\varepsilon^* = \varepsilon' - j\varepsilon''$, (4) the complex dielectric modulus $M^* = M' + jM''$. Indices p and s in the above expressions correspond to parallel and series equivalent circuits, respectively; j is the imaginary unit and ω is the angular frequency. The complex conductivity formalism is also used for analysis of electrical properties: $\sigma^* = j\omega\varepsilon_0\varepsilon^*$ (here ε_0 is the electric constant or dielectric permittivity of free space). Before analysing in detail the IS data, it is recommended that the experimental data is expressed graphically, to reveal their structure, reflecting the studied physical processes. For example, the studied material can be an insulator or a dielectric with leakages, or it can manifest a response, caused by mobile charge in a material with blocking contacts at electrodes. For the entirely blocking contacts, dc current cannot flow through the studied structure, making it difficult to analyse the dielectric response. In the case of conducting materials, the dielectric effects are usually minimal and the most informative approach is the representation in terms of Z^* and M^*. When a non-conducting material is under analysis, the Y and ε formalisms are more acceptable. Nevertheless, the majority of information can be derived from the interpretation of graphical data in all four formalisms regardless of whether a material is a conductor or not. There are different ways of presenting the experimental IS data graphically. When the capacitive effects dominate the inductive ones, it is convenient to draw the dependence of the imaginary part of impedance $-\mathrm{Im}(Z^*)$ (or $-Z''$) versus the real part $\mathrm{Re}(Z^*)$ (or Z'). Represented in such a way, the diagram on the complex plane is called the Nyquist plot. Despite the absence of direct frequency response in this representation, the Nyquist diagram can be very useful to identify different conductivity processes taking place in the studied material. Different areas of a sample associated with their resistance and capacitance are represented by a resistor and a capacitor connected in parallel. The characteristic relaxation time (time constant) corresponding to each 'parallel RC element' is defined by the product of R and C ($\tau = RC$). It is possible to derive various RC elements from the impedance spectra and match them to appropriate areas in a sample. Each parallel RC element of an equivalent circuit causes (ideally) a semicircle in the Nyquist plot from which the components R and C can be extracted. Values of R are obtained by the intercepts on the Z' axis, and the maximum of each semicircle is observed at the frequency of $\omega_m = 1/RC$. Defining ω_m graphically and knowing R, it is possible to calculate the capacitance C.

Semiconducting polycrystalline ceramics like $LiNbO_3$, are inhomogeneous materials in which the resistance of grain boundaries can dominate when IS spectra are recorded and in this case only one semicircle is observed on the Nyquist plot. It this situation some investigators recommend presenting experimental data in terms of complex dielectric modulus $M^* = j\omega C_0 Z^*$ (here $C_0 = \varepsilon_0 S/d$ is the apacitance of an 'empty' capacitor with the electrodes area of S and with distance between them of d). As a result, the response from the grain boundaries as well as from the bulk of grains can be seen. Thus, a peak on the frequency dependence $Z''(\omega)$ corresponds to the element with largest resistance and is equal to the peak of $Z'' = R/2$. A peak on the frequency dependence of the imaginary part of the dielectric modulus $M''(\omega)$ corresponds to an element with minimal capacitance and its value is $M'' = C_0/2\,C$.

It is generally accepted that electrical properties of various materials and structures can be described entirely using an equivalent circuit [55]. We have already mentioned for C–V analysis (see figure 3.9) that at zero or low biases, the studied Si–LiNbO$_3$–Al heterostructures are under the accumulation regime, which allows these heterostructures to be considered as the metal–insulator–metal systems. In other words at low biases the electric properties of the studied heterostructures are affected by the properties of LiNbO$_3$ films and metal contacts, which allows the space-charge areas in Si to be excluded from consideration. Thus, the simplified equivalent circuit of a LiNbO$_3$ film might be represented by those, shown in figure 1.19 (see chapter 1), where R_b and C_b represent resistance and capacitance of the LiNbO$_3$ grain's bulk, whereas R_{gb} and C_{gb} are the resistance and capacitance of the grain boundaries.

The complex impedance, corresponding to the circuit shown in figure 1.19, can be written in the following form:

$$Z^* = \frac{R_b}{1 + j\omega\tau_1} + \frac{R_{gb}}{1 + j\omega\tau_2} \tag{3.47}$$

Here $\tau_1 = R_b C_b$, $\tau_2 = R_{gb} C_{gb}$ are the characteristic times, corresponding to the relaxation processes, j is the the imaginary unit. Equation (3.47) can be modified in the following way:

$$Z^* = \frac{(R_b + R_{gb})(1 + j\omega\tau_o)}{(1 + j\omega\tau_1)(1 + j\omega\tau_2)} \tag{3.48}$$

Here τ_o is given by the following expression:

$$\tau_o = \frac{R_b\tau_2 + R_{gb}\tau_1}{R_b + R_{gb}} \tag{3.49}$$

Taking the logarithm from both sides of equation (3.49) we obtain:

$$\ln(|Z^*|) = \ln(R_b + R_{gb}) + \ln(|1 + j\omega\tau_o|) \\ - \ln(|1 + j\omega\tau_1|) - \ln(|1 + j\omega\tau_2|) \tag{3.50}$$

Therefore, each term in equation (3.50) can be determined independently from a diagram of the coordinates $\ln(|Z^*|) - \ln(\omega)$. The diagrams presenting a frequency dependence of the magnitude and phase of complex impedance in the coordinates $\ln(|Z^*|) - \ln(\omega)$ and $\varphi - \ln(\omega)$ are called the Bode diagram. Each term in formula (3.50) has two frequency limits: the low-frequency case, ($\omega\tau \ll 1$, when $\ln(|1 + j\omega\tau|) = 0$), and the high-frequency case, ($\omega\tau \gg 1$, when $\ln(|1 + j\omega\tau|) = \ln(\omega) + \ln(\tau)$). For these two cases the relaxation time can be derived from the breakpoint on the Bode diagram $\ln(|Z^*|) - \ln(\omega)$. Moreover, it follows from equation (3.50) that $\lim_{\omega \to 0} (|Z^*|) = R_b + R_{gb}$, which allows the total resistance $R_b + R_{gb}$ to be determined from the intercept of a horizontal part of the graph $\ln(|Z^*|) - \ln(\omega)$ with the horizontal axis. In addition, in the disordered materials there is normally a distribution of characteristic time rather than a single relaxation time τ. This leads to deformations in the Nyquist plot, which is not an ideal semicircle in this case, and the complex impedance can be expressed in a way similar to equation (3.47):

$$Z^* = \frac{R_b}{1 + (j\omega\tau_1)^{1-\alpha_1}} + \frac{R_{gb}}{1 + (j\omega\tau_2)^{1-\alpha_2}} \qquad (3.51)$$

Here α_i is the coefficient ($0 \leqslant \alpha_i \leqslant 1$), describing the width of the distribution spectrum of the relaxation time [56] and which can be derived from the slope of a linear section in the Bode diagram $\ln(|Z^*|) - \ln(\omega)$. For comparative analysis of IS spectra we have fabricated the (001)Si–LiNbO$_3$ heterostructures by IBS and RFMS methods under the optimal technological regimes listed in table 2.6 (see chapter 2) in the absence of the ion assist effect.

Figure 3.37 shows the Nyquist and Bode diagrams for the measurement of (001) Si–LiNbO$_3$ heterostructures, recorded at 300 K and fabricated by the RFMS method. The theoretical curves, calculated using equation (3.51) with parameters, obtained through the approximation of experimental data by the the method of least squares are also shown in figure 3.36. Similar diagrams for Si–LiNbO$_3$ hetero-structures, fabricated by IBS method are shown in figure 3.37.

As can be seen from figures 3.36 and 3.37, the Nyquist diagrams are the semicircles, as was expected taking into account the fact that the studied films are the conducting ones. The facts that for both deposition methods semicircles on the Nyquist diagram are elongated and the presence of the breakpoint in the Bode diagram (see figures 3.36 and 3.37) indicate the relaxation time distribution in the films. Analysis of the diagrams shown in figures 3.36 and 3.37 is summarized in table 3.8.

Many authors stressed that the 'slow' relaxation processes (with relatively high characteristic time τ) correspond to the processes at the grain boundaries or at the interfaces, whereas the processes occurring in the bulk of grains cause the 'fast' processes [54, 55]. Thus, presumably, in the studied heterostructures the processes with relaxation time of τ_1 (see table 3.8) take place at the grain boundaries between polycrystalline grains, and the characteristic time τ_2 characterizes the processes, influenced by the bulk properties of LiNbO$_3$ grains. As follows from table 3.8 the heterostructures, fabricated by IBS, manifest the larger dispersion of characteristic times (larger α_1) associated with grain boundaries than those observed for the heterostructures obtained by RFMS. This fact indicates that LiNbO$_3$ films, deposited by the IBS method, demonstrate a higher degree of structural disorder than films fabricated by RFMS method. According to [14], capacitance of grain boundaries in polycrystalline materials similar to the Schottky barrier capacitance correlates with concentration of CLC N_t and the inter-crystalline potential barrier height φ_b as $C_{gb} \propto (N_t/\varphi_b)^{1/2}$. Earlier in this chapter it was demonstrated that the concentration of N_t, apparently attributed to the defects, is considerably higher in LiNbO$_3$ films fabricated by IBS compared to those deposited by RFMS. This accounts for the difference in experimental relaxation times and capacitances C_{gb} presented in table 3.8. For a deeper understanding of the relaxation processes in thin

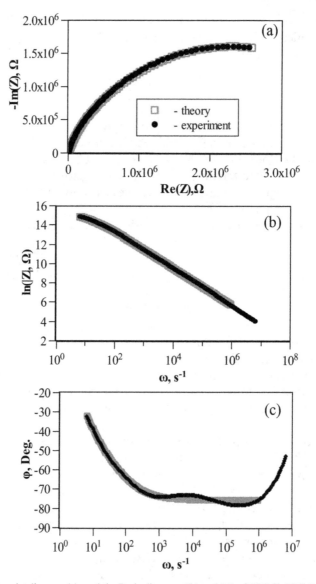

Figure 3.36. The Nyquist diagram (a), and the Bode diagram ((b) and (c)) of (001)Si–LiNbO₃ heterostructures grown by RFMS.

LiNbO$_3$ films a detailed study of the mechanisms of dielectric relaxation and ac conductivity is required.

If an alternating electric field is applied to a material, there is normally a delay between polarization and field. Usually, a dielectric is heated up in the alternating electric field. The part of total energy converted into heat is called the dielectric losses. This is a sum of losses attributed to currents associated with direct voltage and losses, triggered by the active component of the displacement current. The dielectric loss tangent is an important characteristic of dielectric losses [57]:

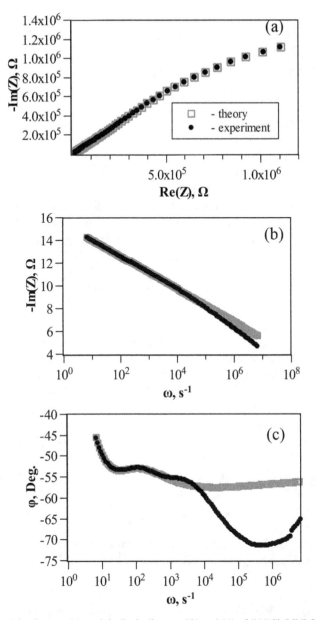

Figure 3.37. The Nyquist diagram (a), and the Bode diagram ((b) and (c)) of (001)Si–LiNbO$_3$ heterostructures, grown by IBS.

Table 3.8. Results of the impedance spectra analysis for (001)Si–LiNbO$_3$, heterostructures, fabricated by two methods.

Fabrication method	R_{gb}, Ohm	R_b, Ohm	τ_1, s	τ_2, s	C_{gb}, F	C_b, F	α_1	α_2
IBS	5.0×10^6	4.0×10^4	0.6	3×10^{-4}	1.2×10^{-7}	7.0×10^{-8}	0.3	0.1
RFMS	4.0×10^6	3.4×10^3	0.1	1×10^{-4}	2.5×10^{-8}	2.9×10^{-8}	0.17	0.2

$$\mathrm{tg}(\delta) = \frac{\varepsilon''}{\varepsilon'} \tag{3.52}$$

Here ε' and ε'' are the real and imaginary parts of complex dielectric permittivity:

$$\varepsilon^* = \varepsilon' - j\varepsilon'' \tag{3.53}$$

The character of dielectric response depends on relaxation mechanisms occurring in the studied material. Polarization mechanisms, caused by the atomic electron polarization or the polarization of an ionic lattice, react almost instantly to the electric field change, contributing only to the real part of ε^*. On the other hand, molecular dipoles, charge defects and carriers taking part in the hopping conductivity (electrons, polarons and ions) trigger the lag in its response on the allied alternating field, contributing to the appearance of imaginary part of ε^*.

In the case of dipole (Debye) polarization, the complex dielectric permittivity can be described by the following expression [56]:

$$\varepsilon^* = \varepsilon_\infty + \frac{\varepsilon_s - \varepsilon_\infty}{1 + (j\omega\tau)^{1-\alpha}} - \frac{j\sigma_{dc}}{\varepsilon_0\omega} \tag{3.54}$$

where ε_s and ε_∞ are the static and high frequency dielectric permittivity, respectively, ε_0 the electric constant, ω is the angular frequency of testing signal, τ the characteristic time for a given process, α the coefficient, characterizing the width of the characteristic times distribution ($0 \leqslant \alpha \leqslant 1$), σ_{dc} the electrical dc conductivity of a sample and j is the complex unit. Usually the second term in equation (3.54) describes the relaxation losses, whereas the third term deals with losses attributed to the low frequency conductivity through the sample.

In the case of charge polarization, when various kinds of charges take part in the hopping conductivity, the difference in charge transport mechanisms can be revealed through the analysis of frequency dependence of electrical conductivity, which in the most general case can be expressed in the following way [14]:

$$\sigma(\omega, T) = \sigma_{dc}(T) + \sigma_{ac}(\omega, T) \tag{3.55}$$

Here σ_{dc} and σ_{ac} are the dc and ac conductivities, respectively. In the disordered systems ac conductivity σ_{ac} is described by the 'universal law' [58]:

$$\sigma_{ac}(\omega, T) = A(T)\omega^s \tag{3.56}$$

Here $A(T)$ is the frequency-independent parameter, ω the angular frequency, and s is the exponent which depends on the particular conductivity mechanism, and usually lies in the range $0.4 < s < 1$. To explain the frequency dependence expressed by equation (3.56) and observed for a wide class of materials, various models were proposed [59–63]. The quantum-mechanical tunneling (QMT) model [59] implies the absence of lattice distortions, associated with motion of charges, taking part in ac conductivity and the frequency exponent s in equation (3.56) does not depend on temperature in the framework of this model. However, this model predicts the following frequency dependence $s(\omega)$:

$$s = 1 - \frac{4}{\ln(1/\omega\tau_0)} \qquad (3.57)$$

Here ω is the angular frequency of testing signal and τ_0 the characteristic relaxation time (the inverse value to the phonon frequency) which within this model is taken as $\tau_0 = 10^{-13}$ s. On the other hand the correlated barrier-hopping (CBH) model, proposed by Elliott [60, 61], considers only the ccorrelated hopping of bipolarons and predicts not only frequency but also temperature dependence of the frequency exponent s. In the framework of the CBH model this parameter decreases with temperature as follows:

$$s = 1 - \frac{6kT}{W_H + kT \ln(\omega\tau_0)} \qquad (3.58)$$

Here k is Boltzmann's constant, τ_0 is the characteristic relaxation time corresponding to the phonon frequency and W_H is the maximum height of a potential barrier that is overcome by carriers. By contrast, the small-polaron tunneling model (SPT model) predicts that the exponent s rises with temperature according to the following law [61]:

$$s = 1 - \frac{4}{\ln(1/\omega\tau_0) - W_H/kT} \qquad (3.59)$$

Therefore, the study of temperature and frequency dependence of s in equation (3.59) is crucial for understanding conduction mechanisms in the studied heterostructures.

Figures 3.38 and 3.39 show the frequency dependences of the dielectric loss tangent for Si–LiNbO$_3$ heterostructures, fabricated by RFMS and IBS methods, respectively.

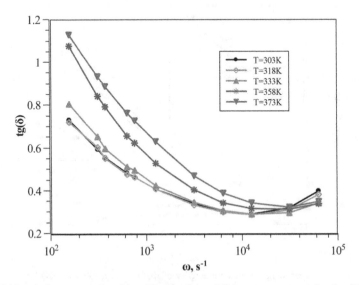

Figure 3.38. Dielectric losses as a function of frequency at different temperatures for the (001)Si–LiNbO$_3$ heterostructures, fabricated by RFMS.

This kind of frequency dependence suggests that the losses in the studied heterostructures result from conductivity rather than that they have a Debye nature. Also, the losses in $LiNbO_3$ films, deposited by IBS are considerably higher than those in the films synthesized by RFMS because of their larger conductivity. For a detailed analysis of the charge transport under alternating voltage we have studied the frequency dependence of conductivity in the temperature range of 290–420 K [47, 64]. Figure 3.40 demonstrates the typical frequency dependence of conductivity of $LiNbO_3$ films fabricated by the RFMS technique.

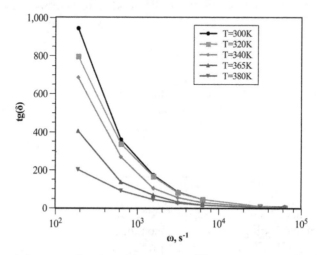

Figure 3.39. Dielectric losses as a function of frequency at different temperatures for the $(001)Si–LiNbO_3$ heterostructures, fabricated by IBS.

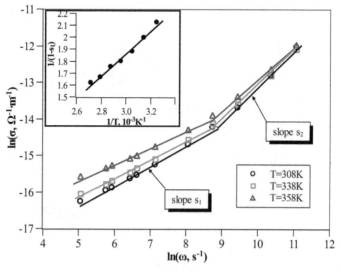

Figure 3.40. Frequency dependence of ac conductivity of the as-grown $(001)Si–LiNbO_3$ heterostructures at different temperatures. The inset shows the variation of the frequency exponent 's_1' in equation (3.56) with temperature.

3-44

Two linear sections can be clearly seen in figure 3.40 with the slopes s_1 and s_2, corresponding to the frequency exponent in equation (3.56) and having different temperature dependence. The fact that s_1 decreases with temperature agrees with equation (3.58) and suggests that in the frequency range of 25–500 Hz the correlated barrier-hopping is the prevalent conductivity mechanism. In this case equation (3.58) can be re-written:

$$\frac{1}{1-s} = \frac{W_H}{6kT} + \frac{1}{6}\ln(\omega\tau_0) \qquad (3.60)$$

In equation (3.60) the maximum barrier height for carriers W_m can be derived graphically from the slope of the linear graph in the coordinates $1/(1-s) - 1/T$ (see inset in figure 3.40) and for the studied films has the magnitude of $W_H = 0.4$ eV [47] which is in good agreement with the results of other authors [65] who studied the hopping conductivity of LiNbO$_3$ single crystals. Moreover, as follows from (3.60), the characteristic time τ_0 can be determined from the intercept of a linear graph $1/(1 - s)$ versus $1/T$ (see inset in figure 3.40). Obtained in this way the relaxation time $\tau_0 = 1.4 \times 10^{-4}$ s is in agreement with the relaxation time τ_2, determined earlier from the IS analysis (see table 3.7) and with the value, reported in [47]. This indicates the relaxation processes caused by bulk properties of grains in LiNbO$_3$ films rather than grain boundaries.

Another informative parameter in equation (3.56) is the coefficient A which is expressed in the following way [61]:

$$A = \frac{D^2 q^{12}}{24\pi^3(\varepsilon\varepsilon_0)^5(kT)^6} \qquad (3.61)$$

Here q is the elementary charge, ε and ε_0 are the dielectric permittivity of a material and the electric constant, respectively, D is the bulk density of CLC in the material. The coefficient A can be determined by extrapolating the frequency dependence of conductivity in the coordinates $\ln(\sigma_{ac})$ versus $\ln(\omega)$ if $\ln(\omega)\rightarrow 0$. After that, it is possible to determine D from A using equation (3.61). Accepting dielectric permittivity as $\varepsilon = 28$ [26], we obtain the density of CLC $D = 7 \times 10^{18}$ cm^{-3}, which correlates with the value, determined earlier from I–V analysis and with the result, reported in [35]. As regards the second linear section on the graph $\ln(\sigma_{ac}) - \ln(\omega)$ with the slope s_2 (see figure 3.40), this stems from the second term in equation (3.60), which dominates at high frequency range ($f > 500$ Hz), when $W_H/kT < \ln(\omega\tau_0)$.

Figure 3.41 shows the frequency dependence of conductivity for the (001)Si–LiNbO$_3$ heterostructures, fabricated by the IBS method.

As in the case of heterostructures grown by RFMS, the frequency dependence of conductivity is linearized in the coordinates $\ln(\sigma_{ac}) - \ln(\omega)$ in total agreement with formula (3.56). The temperature dependence of the frequency exponent s also corresponds to the CBH model (see inset in figure 3.41). Applying the same analysis based on equation (3.60) we come to the value of barrier height $W_H = 1.7$ eV [64], which is exactly equal to the energy of traps in the band gap of LiNbO$_3$, determined in this chapter from the analysis of conductivity mechanisms under dc voltage.

Figure 3.42 demonstrates good agreement between the experimental and theoretical frequency dependence of conductivity, calculated in the framework of the CBH model for (001)Si–LiNbO$_3$ heterostructures.

It follows from figure 3.42 that the correlated barrier-hopping is the predominant charge transport mechanism in the studied LiNbO$_3$ films in the discussed ranges of frequencies and temperatures.

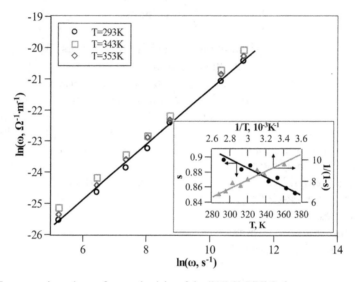

Figure 3.41. Frequency dependence of ac conductivity of the (001) Si–LiNbO$_3$ heterostructures, grown by the IBS method at different temperatures. The inset shows the variation of the frequency exponent 's' with temperature.

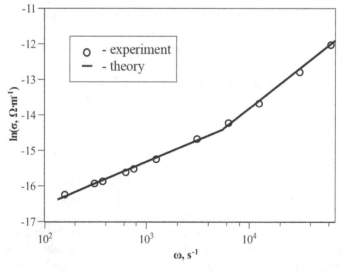

Figure 3.42. Frequency dependence of ac conductivity of the (001)Si–LiNbO$_3$ heterostructures at $T = 300$ K temperature (circles—experiment, line—theoretical conductivity in the framework of the CBH model).

Table 3.9. Basic electrical properties of thin $LiNbO_3$ films, deposited by RFMS and IBS techniques compared to data, reported in literature.

Property	Result	Literature data	Notes	Source
Band gap E_g, eV	4.2 (RFMS)	3.97		[66]
		4.43		[67]
		4.70		[68]
Resistivity ρ, Ohm·cm	1×10^9 (RFMS)	1×10^{10}		[69],[48]
	3×10^9 (IBS)	2×10^9		[6]
Conduction mechanism	DC conductivity:	Hopping electronic	Activation energy E_a = 0.3 eV	[6]
	• Low electric fields ($0 < E < 2$ kV cm^{-1}) and room temperature—hopping conductivity over CLC in the band gap of $LiNbO_3$. Activation energy 0.14 eV.	Hopping polaronic	E_a = 0.5 eV at $T > 420$ K and E_a = 0.067 eV at $T < 420$ K	[70] [36]
	• Moderate fields ($2 < E < 30$ kV cm^{-1})—the Fowler–Nordheim tunneling via CLC in the band gap of $LiNbO_3$ with energy of E_t = 1.7 eV.	Hopping small polarons conductivity	Average energy of 'hops' W = 0.051 eV, 'hopping' between Nb^{5+} and Nb^{4+} ionic states with average energy of W = 0.51 eV at T = 450 K. W = 0.39 eV at T = 450 K (decreases with annealing in vacuum)	[71] [65]
	• Strong fields ($E > 30$ kV cm^{-1})—the thermally-assisted tunneling through the inter-granular potential barrier φ_b = 0.7 eV.	SCLC, Poole–Frenkel emission		[72],[73]
		Richardson–Schottky emission		[48, 74, 75]
		Ionic	Li ions with activation energy of E_a = 0.5 eV	[76]
	AC conductivity: the correlated barrier-hopping conductivity (W_H = 0.4 eV). Frequency dependence obeys the 'universal' power law $\sigma_{ac}(\omega, T) = A(T)\omega^s$		Conductivity $\sigma = 5 \times 10^{-8}$ S cm^{-1} with activation energy of E_a = 0.57 eV	[77]
	28 (RFMS)	27.9		[3],[4]

(Continued)

Property	This work	Value	Material	Reference
Dielectric permittivity ε	29 (IBS)	8		[70]
		29.3		[69]
		46		[3]
		50		[78]
Remnant polarization P_r, μC cm^{-2}	14.0 (RFMS)	71	Bulk LiNbO$_3$	[79]
	11.7 (IBS)	18		[3]
		2.74		[48]
		1.2		[80]
Coercive field E_c, kV cm^{-1}	44.0 (RFMS)	40	Bulk LiNbO$_3$	[81]
	29 (IBS)	115		[82]
		35.23		[29]
		170		[48]
		120		[80]

To sum up, in table 3.9 we summarize the basic electrical properties of LiNbO$_3$ films, synthesized without the ion assist effect compared to the properties, reported in literature.

Summary and emphasis

1. *I–V* and *C–V* analysis revealed that the Si–LiNbO$_3$–Al heterostructures are similar to the metal–insulator–metal system with a conducting dielectric. Basic electrical properties of LiNbO$_3$ films, deposited by both RFMS and IBS methods onto Si substrates were determined. Regardless of the conductivity type of Si wafers, the positive oxide charge with density of $Q_{ef} = 3.8 \times 10^{-8}$ C cm^{-2} is formed, restricting the functionality of LiNbO$_3$-based heterostructures. Concentration of the centers of localized charge depends on the deposition method and maximal for LiNbO$_3$ films synthesized by IBS ($N_t \sim 3 \times 10^{19}$ cm^{-3}) compared to the films, deposited by the RFMS technique ($N_t = 7 \times 10^{17}$ cm^{-3}).
2. The LiNbO$_3$ films, fabricated by both methods without the ion assist effect demonstrate ferroelectric properties. Remnant polarization in the synthesized films is considerably lower than that in bulk lithium niobate, which apparently is caused by the arbitrary grain orientation, limiting application of fabricated heterostructures in the memory units. The built-in field in deposited LiNbO$_3$ films results from the formation of positive oxide charge in the growing process.
3. It has been demonstrated, that charge transport in Si–LiNbO$_3$–Al heterostructures is influenced by the following mechanisms:
   - At low electric fields ($0 < E < 2$ kV cm^{-1}) and room temperature— hopping conductivity over CLC in the band gap of LiNbO$_3$ with activation energy of 0.14 eV.
   - At moderate fields ($2 < E < 30$ kV cm^{-1})—the Schottky emission and the Fowler–Nordheim tunneling, stemmed from CLC, distributed in the band gap of LiNbO$_3$ with energy of $E_t = 1.7$ eV below the bottom of conduction band.
   - At strong electric fields ($E > 30$ kV cm^{-1}) conductivity occurs due to the thermally-assisted tunneling through the inter-granular barrier of $\varphi_b = 0.7$ eV.
4. For the first time the detailed band diagram of the Si–LiNbO$_3$ heterostructure has been designed with all required parameters. It was established, that the band offsets are affected by the high density of interface states at the Si/LiNbO$_3$ heterojunction rather than the electron affinities of contacting materials. The band gap of deposited LiNbO$_3$ films ($E_g = 4.2$ eV) is close to those for bulk lithium niobate.
5. For the first time we have found that during the deposition of LiNbO$_3$ films onto Si substrates shallow donors are generated at the Si/LiNbO$_3$ interface with concentration of $N_d = 5 \times 10^{17}$ cm^{-3} which declines exponentially in the

substrate with coordinate. Apparently, this phenomenon is caused by diffusion of Li atoms in Si substrate during film deposition.

6. With the use of the impedance spectroscopy method it has been demonstrated, that dielectric losses in Si–LiNbO$_3$ heterosystem are caused by the charge relaxation. We revealed two relaxation processes: the 'fast' relaxation (in the bulk of polycrystalline grains) with the characteristic time of $\tau = 1 \times 10^{-4}$ s, and the 'low' one (at grain boundaries) with relaxation time of order $\tau = 0.1$ s.

7. AC conductivity in LiNbO$_3$ films is described in the framework of the correlated barrier-hopping model and obeys the 'universal' power law $\sigma_{ac}(\omega, T) = A(T)\omega^s$. The maximum potential barrier height overcomed by carriers in LiNbO$_3$ films, deposited by RFMS, is found to be $W_H = 0.4$ eV, which is considerably smaller than that for the films, synthesized by IBS method ($W_H = 1.7$ eV).

References

[1] Knack S, Weber J, Lemke H and Riemann H 2002 Copper–hydrogen complexes in silicon *Phys. Rev.* B **65** 165203

[2] Sze S M and Kwok K N 2006 *Physics of Semiconductor Devices* (New York: Wiley)

[3] Simões A Z, Zaghete M A, Stojanovic B D, Riccardi C S, Ries A, Gonzalez A H and Varela J A 2003 LiNbO$_3$ thin films prepared through polymeric precursor method *Mater. Lett.* **57** 2333–9

[4] Choi S-W, Choi Y-S, Lim D-G, moon S-I, Kim S-H, Jang B-S and Yi J 2000 Effect of RTA Treatment on LiNbO$_3$ MFS memory capacitors *Korean J. Ceram.* **6** 138–42

[5] Nassau K, Levinstein H J and Loiacono G M 1966 Ferroelectric lithium niobate. 2. Preparation of single domain crystals *J. Phys. Chem. Solids* **27** 989–96

[6] Shandilya S, Tomar M, Sreenivas K and Gupta V 2009 Purely hopping conduction in *c*-axis oriented LiNbO$_3$ thin films *J. Appl. Phys.* **105** 94105

[7] Strikha V I 1974 *Theoretical Basis of the Operation of Metal - Semiconductor Contacts* (Kiev: Naukova Dumka) [in Russian]

[8] Strikha V I and Buzanova E V 1987 *Physical Basis of the Reability of Metal-Semiconductor Cotacts in Integrated Electronics* (Moscow: Radio and Svyaz')

[9] Maissel L I and Glang R 1970 *Handbook of Thin Film Technology* (New York: McGraw-Hill)

[10] Mott N F 1969 Conduction in non-crystalline materials *Philos. Mag.* **19** 835–52

[11] Shklovskii B I and Efros A L 1984 *Electronic Properties of Doped Semiconductors* vol 45 (Berlin, Heidelberg: Springer)

[12] Iyevlev V, Kostyuchenko A and Sumets M 2011 Fabricatoin, substructure and properties of LiNbO$_3$ films *Proceedings of SPIE - The International Society for Optical Engineering* **7747** 77471J

[13] Hill R M 1971 Poole-Frenkel conduction in amorphous solids *Philos. Mag.* **23** 59–86

[14] Mott N F and Davies E A 1971 *Electronic Processes in Non-Crystalline Materials* (Oxford: Clarendon)

[15] Mycielski J 1961 Mechanism of impurity conduction in semiconductors *Phys. Rev.* **123** 99–103

[16] Adalashvili D I, Adamia Z A, Lavdovskii K G, Levin I and Shklovskii B I 1988 Negative differential resistance in the hopping conductivity region in silicon *JETP* **47** 390–2

[17] Iyevlev V, Sumets M and Kostyuchenko A 2012 Current-voltage characteristics and impedance spectroscopy of $LiNbO_3$ films grown by RF magnetron sputtering *J. Mater. Sci. Mater. Electron* **23** 913–20

[18] Sawyer C B and Tower C H 1930 Rochelle salt as a dielectric *Phys. Rev.* **35** 269–73

[19] Chan H K, Lam C H and Shin F G 2004 Time-dependent space-charge-limited conduction as a possible origin of the polarization offsets observed in compositionally graded ferro-electric films *J. Appl. Phys.* **95** 2665–71

[20] Zhang J, Tang M H, Tang J X, Yang F, Xu H Y, Zhao W F, Zheng X J, Zhou Y C and He J 2007 Bilayer model of polarization offset of compositionally graded ferroelectric thin films *Appl. Phys. Lett.* **91** 162908

[21] Zheng L, Lin C, Xu W-P and Okuyama M 1996 Vertical drift of P–E hysteresis loop in asymmetric ferroelectric capacitors *J. Appl. Phys.* **79** 8634–7

[22] Misirlioglu I B, Okatan M B and Alpay S P 2010 Asymmetric hysteresis loops and smearing of the dielectric anomaly at the transition temperature due to space charges in ferroelectric thin films *J. Appl. Phys.* **108** 34105

[23] Lohkämper R, Neumann H and Arlt G 1990 Internal bias in acceptor-doped $BaTiO_3$ ceramics: Numerical evaluation of increase and decrease *J. Appl. Phys.* **68** 4220–4

[24] Pike G E, Warren W L, Dimos D, Tuttle B A, Ramesh R, Lee J, Keramidas V G and Evans J T 1995 Voltage offsets in (Pb,La)(Zr,Ti)O_3 thin films *Appl. Phys. Lett.* **66** 484–6

[25] Bentarzi H 2011 *Transport in Metal-Oxide-Semiconductor Structures* (Berlin, Heidelberg: Springer)

[26] Iyevlev V, Kostyuchenko A, Sumets M and Vakhtel V 2011 Electrical and structural properties of $LiNbO_3$ films, grown by RF magnetron sputtering *J. Mater. Sci. Mater. Electron* **22** 1258–63

[27] Tagantsev A K and Gerra G 2006 Interface-induced phenomena in polarization response of ferroelectric thin films *J. Appl. Phys.* **100** 51607

[28] Ievlev V, Shur V, Sumets M and Kostyuchenko A 2013 Electrical properties and local domain structure of $LiNbO_3$ thin film grown by ion beam sputtering method *Acta Metall. Sin. (English Lett.)* **26** 630–4

[29] Zhao J P, Liu X R and Qiang L S 2007 Preparation and characterization of $LiNbO_3$ thin films derived from metal carboxylate gels *Key Eng. Mater.* **336–338** 213–6

[30] Waser R and Klee M 1992 Theory of conduction and breakdown in perovskite thin films *Integr. Ferroelectr.* **2** 23–40

[31] Damjanovic D 1998 Ferroelectric, dielectric and piezoelectric properties of ferroelectric thin films and ceramics *Reports Prog. Phys.* **61** 1267–324

[32] Scott J F, Araujo C A, Melnick B M, McMillan L D and Zuleeg R 1991 Quantitative measurement of space-charge effects in lead zirconate-titanate memories *J. Appl. Phys.* **70** 382–8

[33] Duiker H M, Beale P D, Scott J F, Paz de Araujo C A, Melnick B M, Cuchiaro J D and McMillan L D 1990 Fatigue and switching in ferroelectric memories: Theory and experiment *J. Appl. Phys.* **68** 5783–91

[34] Ievlev V, Sumets M and Kostyuchenko A 2013 Conduction mechanisms in Si-$LiNbO_3$ heterostructures grown by ion-beam sputtering method *J. Mater. Sci.* **48** 1562–70

[35] Ievlev V M, Sumets M P and Kostyuchenko A V 2012 Effect of thermal annealing on electrical properties of Si-LiNbO$_3$ *Mater. Sci. Forum* **700** 53–7

[36] Dhar A, Singh N, Singh R K and Singh R 2013 Low temperature dc electrical conduction in reduced lithium niobate single crystals *J. Phys. Chem. Solids* **74** 146–51

[37] Padovani F A and Stratton R 1966 Field and thermionic-field emission in Schottky barriers *Solid. State. Electron.* **9** 695–707

[38] Anderson R L 1962 Experiments on Ge-GaAs heterojunctions *Solid. State. Electron.* **5** 341–51

[39] Yang W-C, Rodriguez B J, Gruverman A and Nemanich R J 2004 Polarization-dependent electron affinity of LiNbO$_3$ surfaces *Appl. Phys. Lett.* **85** 2316–8

[40] Tersoff J 1984 Theory of semiconductor heterojunctions: The role of quantum dipoles *Phys. Rev. B* **30** 4874–7

[41] Mönch W 1990 Role of virtual gap states and defects in metal-semiconductor contacts *Electronic Structure of Metal-Semiconductor Contacts* ed W Mönch (Dordrecht: Springer), pp 224–7

[42] Ievlev V, Sumets M, Kostyuchenko A, Ovchinnikov O, Vakhtel V and Kannykin S 2013 Band diagram of the Si-LiNbO$_3$ heterostructures grown by radio-frequency magnetron sputtering *Thin Solid Films* **542** 289–94

[43] Fox M 2010 *Optical Properties of Solids* (New York: Oxford University Press)

[44] Thierfelder C, Sanna S, Schindlmayr A and Schmidt W G 2010 Do we know the band gap of lithium niobate? *Phys. Status Solidi.* **7** 362–5

[45] Bhatt R, Bhaumik I, Ganesamoorthy S, Karnal A K, Swami M K, Patel H S and Gupta P K 2012 Urbach tail and bandgap analysis in near stoichiometric LiNbO$_3$ crystals *Phys. Status Solidi.* **209** 176–80

[46] Reik H G and Heese D 1967 Frequency dependence of the electrical conductivity of small polarons for high and low temperatures *J. Phys. Chem. Solids* **28** 581–96

[47] Ievlev V, Sumets M, Kostyuchenko A and Bezryadin N 2013 Dielectric losses and ac conductivity of Si-LiNbO$_3$ heterostructures grown by the RF magnetron sputtering method *J. Mater. Sci. Mater. Electron.* **24** 1651–7

[48] Lim D, Jang B, moon S, Won C and Yi J 2001 Characteristics of LiNbO$_3$ memory capacitors fabricated using a low thermal budget process *Solid. State. Electron.* **45** 1159–63

[49] Gosele U M 1988 Fast diffusion in semiconductors *Annu. Rev. Mater. Sci.* **18** 257–82

[50] Yoshimura K, Suzuki J, Sekine K and Takamura T 2007 Measurement of the diffusion rate of Li in silicon by the use of bipolar cells *J. Power Sourc.* **174** 653–7

[51] Simmons J G 1965 Richardson-Schottky effect in solids *Phys. Rev. Lett.* **15** 967–8

[52] Zubko P, Jung D J and Scott J F 2006 Space charge effects in ferroelectric thin films *J. Appl. Phys.* **100** 114112

[53] Macdonald J R 1992 Impedance spectroscopy *Ann. Biomed. Eng.* **20** 289–305

[54] Irvine J T S, Sinclair D C and West A R 1990 Electroceramics: Characterization by impedance spectroscopy *Adv. Mater.* **2** 132–8

[55] Hodge I M, Ingram M D and West A R 1976 Impedance and modulus spectroscopy of polycrystalline solid electrolytes *J. Electroanal. Chem. Interfacial Electrochem.* **74** 125–43

[56] Brown W F 1956 Dielectrics *Encyclopedia of Physics* (Berlin: Springer), pp 1–154

[57] Zheludev I S 1971 Physics of Crystalline Dielectrics (New York: Plenum)

[58] Jonscher A K 1977 The 'universal' dielectric response *Nature* **267** 673–9

[59] Austin I G G and Mott N F F 1969 Polarons in crystalline and non-crystalline materials *Adv. Phys.* **18** 41–102

[60] Elliott S R 1977 A theory of a.c. conduction in chalcogenide glasses *Philos. Mag.* **36** 1291–304

[61] Elliott S R 1987 A.c. conduction in amorphous chalcogenide and pnictide semiconductors *Adv. Phys.* **36** 135–217

[62] Pollak M and Pike G E 1972 AC conductivity of glasses *Phys. Rev. Lett.* **28** 1449–51

[63] Long A R 1982 Frequency-dependent loss in amorphous semiconductors *Adv. Phys.* **31** 553–637

[64] Ievlev V, Sumets M and Kostuchenko A 2013 Electrical conductivity of the Si-LiNbO$_3$ heterostructures grown by ion sputtering method *Proc. SPIE* **8770** 8770M1

[65] Akhmadullin I S, Golenishchev-Kutuzov V A, Migachev S A and Mironov S P 1998 Low-temperature electrical conductivity of congruent lithium niobate crystals *Phys. Solid State* **40** 1190–2

[66] Fakhri M A, Al-Douri Y, Hashim U, Salim E T, Prakash D and Verma K D 2015 Optical investigation of nanophotonic lithium niobate-based optical waveguide *Appl. Phys. B Lasers Opt.* **121** 107–16

[67] Shandilya S, Sharma A, Tomar M and Gupta V 2012 Optical properties of the *c*-axis oriented LiNbO$_3$ thin film *Thin Solid Films* **520** 2142–6

[68] Satapathy S, Mukherjee C, Shaktawat T, Gupta P K and Sathe V G 2012 Blue shift of optical band-gap in LiNbO$_3$ thin films deposited by sol–gel technique *Thin Solid Films* **520** 6510–4

[69] Gupta V, Bhattacharya P, Yuzyuk Y I, Katiyar R S, Tomar M and Sreenivas K 2004 Growth and characterization of *c*-axis oriented LiNbO$_3$ film on a transparent conducting Al: ZnO inter-layer on Si *J. Mater. Res.* **19** 2235–9

[70] Easwaran N, Balasubramanian C, Narayandass S A K and Mangalaraj D 1992 Dielectric and AC conduction properties of thermally evaporated lithium niobate thin films *Phys. Status Solidi.* **129** 443–51

[71] Dhar A and Mansingh A 1990 Polaronic hopping conduction in reduced lithium niobate single crystals *Philos. Mag. Part* B **61** 1033–42

[72] Hao L Z, Zhu J and Li Y R 2011 Integration between LiNbO$_3$ ferroelectric film and AlGaN/GaN system *Mater. Sci. Forum* **687** 303–8

[73] Joshi V, Roy D and Mecartney M L 1995 Nonlinear conduction in textured and non textured lithium niobate thin films *Integr. Ferroelectr.* **6** 321–7

[74] Hao L-Z, Liu Y-J, Zhu J, Lei H-W, Liu Y-Y, Tang Z-Y, Zhang Y, Zhang W-L and Li Y-R 2011 Rectifying the current-voltage characteristics of a LiNbO$_3$ film/GaN heterojunction *Chinese Phys. Lett.* **28** 107703

[75] Guo S M, Zhao Y G, Xiong C M and Lang P L 2006 Rectifying I-V characteristic of LiNbO$_3$/Nb-doped SrTiO$_3$ heterojunction *Appl. Phys. Lett.* **89** 223506

[76] Kim S H, Lee S J, Kim J P, Chae B G, Yang Y S and Jang M 1998 Low-frequency dielectric dispersion and Raman spectroscopy of amorphous LiNbO$_3$ *J. Koreal Phys. Soc.* **32** 830–3

[77] Can N, Ashrit P V, Bader G, Girouard F and Truong V 1994 Electrical and optical properties of Li-doped LiBO$_2$ and LiNbO$_3$ films *J. Appl. Phys.* **76** 4327–31

[78] Edon V, Rèmiens D and Saada S 2009 Structural, electrical and piezoelectric properties of LiNbO$_3$ thin films for surface acoustic wave resonators applications *Appl. Surf. Sci.* **256** 1455–60

[79] Wemple S H, DiDomenico M and Camlibel I 1968 Relationship between linear and quadratic electro-optic coefficiens in $LiNbO_3$, $LiTaO_3$, and other oxygen-octahedra ferroelectrics based on direct measurement of spontaneous polarization *Appl. Phys. Lett.* **12** 209–11

[80] Kim K-H, Lee S-W, Lyu J-S, Kim B-W and Yoo H-J 1998 Properties of lithium niobate thin films by RF magnetron sputtering with wafer target *J. Korean Phys. Soc.* **32** 1508—1512

[81] Gopalan V, Mitchell T E, Furukawa Y and Kitamura K 1998 The role of nonstoichiometry in 180° domain switching of $LiNbO_3$ crystals *Appl. Phys. Lett.* **72** 1981

[82] Kim Y-S, Jung S-W, Jeong S-H, In Y-I, Kim K-H and No K 2003 Properties of $LiNbO_3$ thin films fabricated by CSD (Chemical Solution Decomposition) method AWAD2003: Asia-Pacific Workshop on Fundamental and Application of Advanced Semiconductor Devices *Technical report of IEICE. SDM* vol 103 pp 33–6

Chapter 4

Effect of sputtering conditions and thermal annealing on electron phenomena in the Si–LiNbO$_3$ heterostructures

It was shown in chapter 3, that a relatively high concentration of centers of localized charge (CLC) and positive oxide charge as well as low remnant polarization in LiNbO$_3$ films, fabricated without the ion assist effect, restrict applications of LiNbO$_3$-based heterostructures. Since technological regimes of RFMS affect the structure and surface morphology of LiNbO$_3$ films, the optimal deposition regimes providing the formation of single phase oriented films with a high degree of crystallinity and minimal surface roughness, were proposed in chapter 2.

Motivated by the lack of systematic investigations on the influence of synthesis conditions on electrical properties of fabricated LiNbO$_3$-based heterostructures in this chapter we focus on studying methods to change these properties through a variety of technological regimes of RFMS followed by thermal annealing. It was demonstrated in chapter 2 and in [1, 2] that fundamental sputtering parameters such as reactive gas pressure and composition, the temperature of a substrate and its relative position with a target are critical for structure, composition and surface morphology of deposited LiNbO$_3$ films. Furthermore, further thermal annealing (TA) leads to recrystallization of the films, disappearance of texture and formation of 'parasitic' phases like LiNb$_3$O$_8$. There is no doubt that all these facts should be reflected in the change of electrical properties of LiNbO$_3$-based heterostructures. For electrical measurements we fabricated films by RFMS and IBS methods onto heated ($T = 550$ °C) silicon wafers of n-type conductivity ($\rho = 4.5$ Ohm cm) under optimal conditions (see chapter 2) *with the ion assist effect* to improve polarization reversal.

As described in chapter 3, electrical properties of fabricated heterostructures were studied by the techniques based on obtaining current voltage (I–V) and high frequency (1 MHz) capacitance–voltage (C–V) characteristics, the Sawyer–Tower

doi:10.1088/978-0-7503-1729-0ch4

method and also based on the tangent loss frequency dependence and impedance spectroscopy (IS). The local domain structure was studied by the piezoresponse force microscopy (PFM) method. The fundamentals of this method can be found in [3].

4.1 Effect of the spatial plasma inhomogeneity, composition and relative target–substrate position on electrical properties of Si–LiNbO$_3$ heterostructures

Films were synthesized by the RFMS and IBS methods under the ion assist effect according to the regimes listed in table 4.1 to study the influence of reactive gas pressure on the film's electrical properties.

Structural properties of heterostructures similar to those listed in table 4.1 were described in detail in chapter 2 and in [1, 4, 5], so we do not analyse them here.

4.1.1 Capacitance–voltage characteristics and ferroelectric properties of Si–LiNbO$_3$–Al heterostructures, fabricated under different regimes

Figure 4.1 shows typical normalized high frequency $C–V$ characteristics of studied Si–LiNbO$_3$–Al heterostructures that are also similar to ones for MIS structures (see chapter 3).

As can be seen from figure 4.1, $C–V$ characteristics of samples LN133 and LN134 are shifted to the left along the horizontal axis which is evidence of positive oxide charge, existing in the LiNbO$_3$ films of all samples. Applying standard methods of $C–V$ analysis we determined the effective density of positive charge and effective density of states in the studied Si–LiNbO$_3$–Al heterostructures [5]. Table 4.2 summarizes these results.

As shown in table 4.2 similar to previous cases when films were deposited without the ion assist effect, in the absence of bias or even at low positive bias ('+' at Al) the studied heterostructures are in the accumulation regime, so they can be analysed as a metal–insulator–metal system. Also, it is important to note that the effective density of CLC in the films, deposited at higher reactive gas pressure (sample LN133), is close to those for the films fabricated at nearly optimal conditions (sample LN134). In chapter 2 and in [1] we demonstrated that the increase of reactive gas pressure is five-fold compared with the optimal one (0.1 Pa) and this leads to intensive bombardment of the film surface by plasma particles, causing formation of defective layers, texture disappearance and creation of a 'parasitic' LiNb$_3$O$_8$ phase. Consequently, the effective positive charge in deposited films is not caused either by formation of the LiNb$_3$O$_8$ phase or by defects, produced by bombardment. Almost the same value of positive charge was observed by other researchers in LiNbO$_3$ films, fabricated by the laser ablation method [6], and it was explained by the formation of lithium vacancies. On the other hand, films deposited in an Ar + O$_2$ reactive gas environment (sample LN135) demonstrate the lowest effective density of positive charge. Thus, the presence of oxygen in a reactive chamber plays a key role in reducing built-in charge. It is generally accepted that mechanical strains affect structural and electrical properties of deposited coatings. Strains in a film can be

Table 4.1. Technological regimes of fabrication of Si–LiNbO$_3$–Al heterostructures by RFMS method.

Sample #	Film thickness, d (µm)	Magnetron power (W)	Reactive gas composition	Reactive gas pressure, P (Pa)	Substrate–target distance (cm)	Film composition and structure
LN133	1	100	Ar	5.0×10^{-1}	5	Polycrystalline two-phase (LiNbO$_3$ and LiNb$_3$O$_8$) films with arbitrary grain orientation
LN134	0.37	100	Ar	1.5×10^{-1}	5	Single-phase polycrystalline <0001> textured LiNbO$_3$ films
LN135	0.60	100	Ar(60%) + O$_2$(40%)	1.5×10^{-1}	5	Single-phase polycrystalline <0001> textured LiNbO$_3$ films

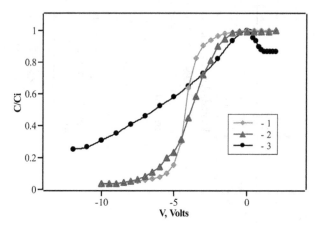

Figure 4.1. Typical C–V characteristics of studied Si–LiNbO$_3$–Al heterostructures at temperature $T = 293$ K. 1—sample LN133, 2—sample LN134, 3—sample LN135.

Table 4.2. Results of C–V analysis for Si–LiNbO$_3$–Al heterostructures, fabricated by RFMS under different regimes.

Sample #	Dielectric constant of a film, ε	Flat band voltage, V_{FB} (V)	Effective charge in a film, Q_{eff} (C cm^{-2})	Effective density of states, N_{eff} (cm^{-2})	Position of a charge centroid (relative to a film surface) in thickness units, d_c/d
LN133	32	−4.8	+9 × 10^{-7}	5.7 × 10^{12}	0.96
LN134	28	−5.6	+8 × 10^{-7}	5.0 × 10^{12}	0.36
LN135	29	−3	+5.1 × 10^{-8}	3.2 × 10^{11}	–

divided into two types: internal and external. Internal strains are triggered by point defects and lattice distortions in a deposited layer and these strains occur inevitably during the RFMS process due to relatively intensive bombardment of a substrate surface by high energetic plasma particles. The crystal lattices mismatch and the difference in linear expansion coefficients at the film–substrate interface is the main source of external strain. According to the classification given in chapter 3, oxide charge is the sum of the interface-trapped charge (Q_{it}), oxide-trapped charge (Q_{ot}), fixed oxide charge (Q_f) and the mobile ionic charge (Q_{mi}) which is immobile at temperatures below 390 °C. Since in our case Q_{eff} does not depend on the substrate type and orientation or on the intensity of bombardment of a film surface, we can exclude Q_{it} from consideration assuming that $Q_{eff} = Q_{ot} + Q_f$. Thus, internal strains are the main source of positive oxide charge observed in the studied LiNbO$_3$ films. It was shown in [7] that technological parameters of RFMS affect the degree of mechanical strains in LiNbO$_3$ films greatly. Specifically, it was recommended that sputtering at pressure of 1.3 Pa should be conducted in a reactive chamber using an Ar(80%) +O$_2$(20%) gas mixture as a reactive environment. Authors of work [7] indicated that these conditions provide the formation of LiNbO$_3$ films with minimal internal strains associated with oxygen vacancies. On the other hand, there is strong evidence that the presence of O$_2$ in the reactive chamber influences plasma properties, increasing concentration of Li ions [8]. Analysis of various models of defects formation suggests that the $Nb_{Li}^{4\bullet}$ antisite positively charged defects (Nb on an Li site) are the most likely source of positive charge in LiNbO$_3$ single crystals [9]. If so, the presence of oxygen reduces the concentration of lithium vacancies (V_{Li}) and consequently the formation of positively charged complexes $Nb_{Li}^{4\bullet}$.

Before we determine the distribution of electrically active defects and impurities in Si substrate and in LiNbO$_3$ films it is interesting to analyse C–V characteristics of sample LN135. As can be clearly seen from figure 4.1, a peak and further modulation of capacitance are observed in the range of positive biases that radically differs from samples LN133 and LN134. This unusual behavior of C–V curves can be explained in the framework of electrical characteristics of isotype heterojunctions. In [10] the authors explained a similar C–V curve for n–n heterojunctions based on the double depletion layer model. The heterojunction in the framework of this model is modeled as two Schottky diodes connected in series back-to-back providing a high density of electronic states exists at the interface, which is equivalent to the occurrence of a metal contact between two barriers (see figure 4.2).

In heterostructures, fabricated in an Ar(60%) + O$_2$(40%) environment by RFMS, an extended intermediate layer with variable composition is formed (see chapter 3 and [4]). Assuming that this layer has high conductivity, which is supported by I–V analysis, it should have a large concentration of CLC. Thus, the presence of this layer can be modeled as a metal contact between two semiconducting materials, as shown in figure 4.2.

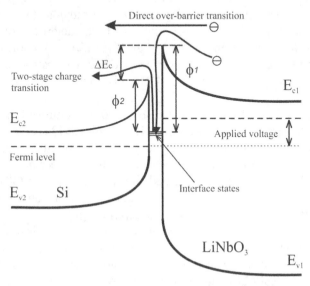

Figure 4.2. Schematic band diagram of an n–n isotype heterojunction and applied voltage as two in series back-to-back Schottky diodes, forming a double depletion layer. The arrows indicate two possible charge transport mechanisms.

From figure 4.2 it follows that, regardless of the polarity of applied voltage, C–V characteristic decreases with bias and it is affected by capacitance with respect to the reverse biased Schottky diode. With regards to electrical conductivity, which will be analysed in the next section, it is limited by the direct over barrier charge transition as well as by the two-stage charge transport, indicated by arrows in figure 4.2. Let us determine the spatial distribution of electrically active defects and impurities in the Si substrates and $LiNbO_3$ films.

In the framework depletion layer approximation, concentration of doping impurities in a semiconductor $N(x)$ can be expressed through the measured high frequency capacitance C by the following expression [11]:

$$N(x) = \pm\frac{2}{q\varepsilon_0\varepsilon_s}\left[\frac{d}{dV}\left(\frac{S}{C}\right)^2\right] \tag{4.1}$$

Where '$-$' represents an n-type semiconductor and '$+$' corresponds to a p-type semiconductor. In equation (4.1) q is the electron charge, $\varepsilon_0 = 8.85 \times 10^{-12}$ F m^{-1} is the dielectric permittivity of vacuum, ε_s is the dielectric permittivity of a material, S the contact area, V the applied voltage. Thus, the slope of the experimental C–V characteristic in $(S/C)^2$–V coordinates is determined by impurity concentration $N(x)$ in the depletion layer. Coordinate x in equation (4.1) is calculated using the measured capacitance of a MIS structure [11]:

$$x = \varepsilon_0\varepsilon_s S\left(\frac{1}{C} - \frac{1}{C_i}\right) \tag{4.2}$$

Here C_i is the capacitance of a dielectric layer (capacitance of the MIS structure in the accumulation regime). As regards the Schottky diodes, a doping profile in a semiconductor can also be determined using equations, similar to (4.1) and (4.2) in the approximation of $1/C_i \to 0$. Figure 4.3 shows C–V characteristics of Si–$LiNbO_3$–Al heterostructures in the depletion regime ('$-$' at Al) in $(S/C)^2$–V coordinates.

Donor distributions, in Si substrates of the studied heterostructures, obtained by graphical integration of experimental C–V curves and using equations (4.1) and (4.2) are shown in figure 4.4.

As follows from figure 4.4 donors are distributed non-uniformly with thickness in substrates of all samples. In samples LN133 and LN134 donor concentration declines with a coordinate and at the depth of 500 nm it reaches the nominal concentration, attributed to n-Si substrates, used in our experiments [5]. It is worth noting, that in heterostructures fabricated in an Ar(60%) + O_2(40%) gas mixture (sample LN135) donor concentration is considerably higher than in those formed in a pure Ar atmosphere and is extended to a depth of up to 1 μm. This result can be explained based on the fact that diffusive molecular oxygen leads to the formation of donor centers in silicon [12]. This process is very sensitive to substrate temperature and oxygen pressure. Molecular oxygen has the diffusion coefficient of $D = 10^{-9}$ cm^2 s^{-1} at 450 °C which is ten-fold of those for atomic oxygen. On the other hand, it was shown in [8] that the presence of oxygen in a reactive chamber affects reactive gas

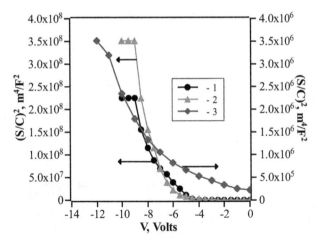

Figure 4.3. C–V characteristics of the studied Si–LiNbO$_3$–Al heterostructures in $(S/C)^2$—V coordinates in the depletion regime ('–' at Al).

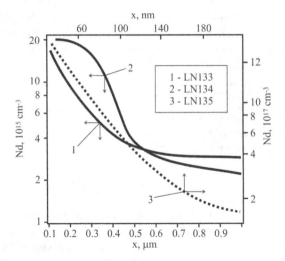

Figure 4.4. Donor concentration profile in silicon substrate for the studied Si–LiNbO$_3$–Al heterostructures.

composition, increasing concentration of Li atoms in a plasma. As a result, Li atoms, being shallow donors in Si [13] with diffusion coefficient of $D = 10^{-11}$ cm^2 s^{-1}, penetrate the substrate and do not form a uniform impurity distribution in full agreement with our experimental results.

The Schottky diode theory states that in the case of uniform donor distribution in a semiconductor with concentration of N_d, capacitance of the Schottky barrier is a linear function of applied depleting voltage in $(S/C)^2$–V coordinates and is described by the following equation [11]:

$$\left(\frac{S}{C}\right)^2 = \frac{2(\varphi_D - V)}{q\varepsilon\varepsilon_0 N_d} \tag{4.3}$$

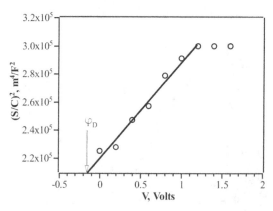

Figure 4.5. Plot of $(S/C)^2$ versus V for sample LN135, associated with the depletion zone in LINbO$_3$ at $T = 295$ K.

Here φ_D is the diffusion potential, ε the dielectric constant of a semiconductor, V the applied voltage. Thus, the donor concentration can be found through the slope of a $(S/C)^2$–V graph and φ_D is defined by the intercept with the horizontal axis.

Figure 4.5 shows C–V characteristic of the LN135 sample at positive biases according to a depletion zone of the LiNbO$_3$ layer (attributed to capacitance of the Schottky barrier in LiNbO$_3$) in $(S/C)^2$–V coordinates.

It follows from figure 4.5 that there is a uniform donor distribution in LiNbO$_3$ film with a concentration of $N_d = 7 \times 10^{17}$ cm^{-3}, that was derived from the slope of the linear part in figure 4.5 with the use of equation (4.3) and taking $\varepsilon = 28$ [14]. Furthermore, the diffusion potential, determined from the same graph is $\varphi_D = 0.3$ V.

We also determined the energy distribution of the surface state's D_{ss} at Si/LiNbO$_3$ interface through the shift of experimental C–V curves relative to the theoretical ones. These distributions in the upper half of the band gap of Si for the studied samples are shown in figure 4.6.

From figure 4.6 it follows that energy distribution does not change significantly with increase of Ar pressure in a reactive chamber and D_{ss} is one order lower than those, reported by other authors [15]. The density of states in heterostructures fabricated in an Ar(60%) + O$_2$(40%) reactive gas mixture (sample LN135) greatly exceeds the D_{ss} of those grown in pure Ar (samples LN133 and LN134). This agrees with the double depletion layer model with high density of states between two Schottky diodes according to which the electrical properties of sample LN135 are described (see figure 4.2).

Films, fabricated in all three regimes, manifested ferroelectric properties [5], which reflected in ferroelectric hysteresis loops, shown in figure 4.7.

The results of analysis of P–E loops shown in figure 4.7 are presented in table 4.3. It follows from these results that the remnant polarization does not depend on the deposition regimes used and it is close to those for single crystal lithium niobate $P_r = 71$ μC cm^{-2} [16].

It is important to note, that the remnant polarization for all samples is considerably higher than those for LiNbO$_3$ films with arbitrary grain orientation,

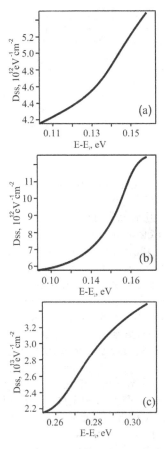

Figure 4.6. Energy distribution of surface states in the upper half of the band gap of Si for the studied Si–LiNbO$_3$–Al heterostructures ((a) sample LN133, (b) sample LN134, (c) sample LN135).

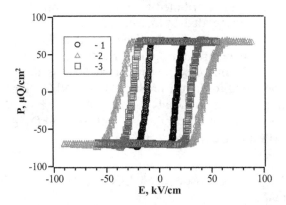

Figure 4.7. *P–E* loops for the studied heterostructures: 1 sample LN133, 2 sample LN134, 3 sample LN135.

Table 4.3. Parameters of *P–E* loops for the studied heterostructures.

Sample #	Remnant polarization, P_r (μC cm^{-2})	Coercive field, E_c (kV cm^{-1})	Vertical shift of *P–E* loops, ΔP_r (μC cm^{-2})	Built-in field, E_b (kV cm^{-1})
LN133	69	14.0	−1	2.8
LN134	69	39.5	−1	2.8
LN135	69	27.3	−1	2.0

grown without the ion assist effect and studied in chapter 3. Apparently, a single-axis <0001> texture in LiNbO$_3$ films, deposited under the ion assist regimes (see table 4.1) promotes the effective polarization reversal in the studied films making their ferroelectric properties close to those for single crystal lithium niobate. The coercive field is minimal for the films fabricated at higher working pressure and, apparently it depends on the ferroelectric domain's dynamics, but this issue is beyond the scope of this monograph and requires additional study. Moreover, like LiNbO$_3$ films with arbitrary orientation of grains (see chapter 3), the textured films studied here also demonstrate built-in field and a slight shift of *P–E* loops along the vertical axis which can be caused by preferable domain orientation. As was discussed in chapter 3, the most likely origin of this built-in field in LiNbO$_3$ films can be the oxide of interface-trapped charge. In fact, as follows from the *C–V* analysis positive oxide charge presented in LiNbO$_3$ films does not depend on reactive gas pressure (samples LN133 and LN134 in table 4.2) and it declines when oxygen presents in the reactive chamber (sample 135). This fact correlates with data, related to the built-in field in table 4.3).

There is some evidence that various defects and space charge effects influence the *P–E* loops greatly [17, 18]. The coercive field in this case can be expressed as:

$$E_c = E_c' - E_{sc} + E_{defect} \tag{4.4}$$

Here E_c' is the coercive field attributed to domain motion, E_{sc} is the space charge field, E_{defect} is the field generated by defects in a ferroelectric film. The minimal coercive field (see table 4.3) for films, fabricated at higher working gas pressure (sample LN133) and having arbitrary grain orientation, is caused by terms E_c' and E_{sc} in equation (4.4). In this case lower electric field is needed for ferroelectric domain reversal compared to the textured <0001> LiNbO$_3$ films (samples LN 134 and LN 135 in table 4.1). Further study requires the analysis of *I–V* characteristics and frequency dependence of ac conductivity of Si–LiNbO$_3$–Al heterostructures, fabricated by RFMS under different regimes.

4.1.2 Current–voltage characteristics of Si–LiNbO$_3$–Al heterostructures, fabricated under different regimes

I–V characteristics of Si–LiNbO$_3$–Al heterostructures in ln *J–V* coordinates at direct biases ('+' at Al) are shown in figure 4.8.

The fast increase with applied voltage sections on *I–V* curves, observed in figure 4.8 is influenced by the barrier properties of Si/LiNbO$_3$ heterojunction, whereas the second sloping sections are attributed to bulk properties of the films (see discussion in chapter 3). It follows from figure 4.8 that barrier properties of structures, fabricated in an Ar environment are not influenced by reactive gas pressure (samples LN133 and LN134), whereas these properties depend on reactive gas composition (sample LN135 in figure 4.8).

First, let us analyse the possible conduction mechanisms in sample LN133 based on the band diagram of the studied heterostructures proposed in chapter 3. At relatively low direct biases the most probable barrier-limited conduction mechanisms are the Fowler–Nordheim tunneling (at low temperatures) and the Richardson–Schottky emission (at higher voltage). The electronic currents corresponding to these mechanisms are indicated as J_{F-N} and J_{R-S} in figure 4.9.

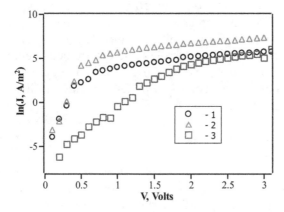

Figure 4.8. Typical *I–V* characteristics Typical *I–V* characteristics of Si–LiNbO$_3$–Al heterostructures, fabricated by RFMS method under different conditions at direct biases ('+' at Al contact) and temperature of $T = 300$ K. 1 sample LN133, 2 sample LN134, 3 sample LN135.

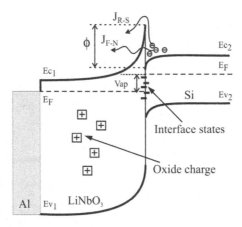

Figure 4.9. Band diagram of sample LN133 at low direct biases V_{ap}. The curved arrows indicate two conduction mechanisms, limited by the potential barrier of ϕ.

It was shown in chapter 1, that in the case of the Fowler–Nordheim tunneling I–V characteristics should be linear in $\ln(J/E^2)$–$1/E$ coordinates where J is the current density and E is the the electric field strength. The slope of this linear graph is determined by the potential barrier height ϕ and can be described by the following expression [19]:

$$B = \frac{8\pi\sqrt{2m^*}\,\phi^{3/2}}{3qh} \tag{4.5}$$

Here, h is Planck's constant, m^* is the electron effective mass and ϕ is the average potential barrier height. Actually, the I–V characteristics of Si–LiNbO$_3$–Al hetero-structures, fabricated at relatively high working pressure (sample LN133) and recorded at low temperatures have a linear section in the Fowler–Nordheim coordinates (see figure 4.10) which suggests tunneling in this temperature range.

Taking into account corrections, caused by polarization effects in a ferroelectric film (see equation (3.41) in chapter 3), the potential barrier height, derived from the slope of the linear part of the I–V curve in figure 4.10 is $\phi = 0.4$ eV, which is less than that for heterostructures similar to LN134 and determined in chapter 3 (the value of conduction band offset ΔE_c). Also, it is seen from figure 4.10 that the tunnel component of conductivity declines with temperature, which is reflected in the shortening of the linear section of I–V curves.

Let us analyse the I–V characteristics of sample LN135 at relatively low biases (low applied electric field). According to [10] in the framework of the double depletion model (two connected in series back-to-back Schottky barriers), I–V characteristics are limited by the two-stage charge transport mechanism (see figure 4.2) and greatly depend on the Schottky barrier heights ϕ_1 and ϕ_2. The current densities through each of the two diodes are described by the following equations:

$$J_1 = J_{s1}(\exp(V_1/V_{o1}) - 1); \quad J_2 = J_{s2}(\exp(V_2/V_{o2}) - 1) \tag{4.6}$$

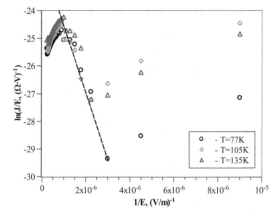

Figure 4.10. I–V characteristics of sample LN133 in the Fowler–Nordheim coordinates. The dash line indicates the Fowler–Nordheim tunneling through the potential barrier of ϕ (current $J_{\mathrm{F–N}}$ in figure 4.9).

Here V_1 and V_2 are the voltage drops on barriers ϕ_1 and ϕ_2, respectively, V_o is the coefficient dependent on a conduction mechanism through the barrier, J_{s1} and J_{s2} are the saturation currents, denoted by the following equations:

$$J_{s1} = A^* T^2 \exp\left(-\frac{q\phi_1}{kT}\right) J_{s2} = A^* T^2 \exp\left(-\frac{q\phi_2}{kT}\right) \qquad (4.7)$$

Here A^* is the effective Richardson constant, k is the Boltzmann constant, T the temperature. Since $J_1 = J_2$ and $V = V_1 + V_2$, we can obtain the following expression for total current density:

$$J_{tot} = \frac{J_{s1}J_{s2}(\exp(qV/kT) - 1)}{J_{s2} + J_{s1}\exp(qV/kT)} \qquad (4.8)$$

the I–V characteristics, described by equation (4.8) have an inflection point, corresponding to the following voltage and current density:

$$V_{inf} = \frac{kT \ln(J_{s2}/J_{s1})}{q} J_{inf} = \frac{1}{2}J_{s1}(J_{s2}/J_{s1} - 1) \qquad (4.9)$$

In our case $J_{s2} \gg J_{s1}$, and $J_{inf} = J_{s2}/2$. Furthermore, as follows from equation (4.7), the potential barrier height ϕ can be found from the linear section of the $\ln(J_s/T^2)$ – $1/T$ graph [4]. The graph for sample LN135 is shown in figure 4.11.

Figure 4.11 shows that the temperature dependence of the saturation current is approximated by a linear law $\ln(J_s/T^2)$ – $1/T$ only at relatively high temperatures (140–300 K). Deviation from a straight line at low temperatures results from the fact that, within this temperature range, the Fowler–Nordheim tunneling rather than the Richarson–Schottky emission affects I–V curves. The Schottky barrier height (at the Si site), determined from the slope of this linear section in figure 4.11 is equal to $\phi_2 = 0.27$ eV.

On the other hand, by substituting equation (4.7) in equation (4.9) for V_{inf} we can obtain the following expression:

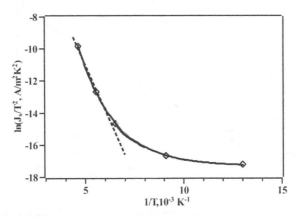

Figure 4.11. Temperature dependence of saturation current J_s for sample LN135 in $\ln(J_s/T^2)$ – $1/T$ coordinates.

$$V_{\text{inf}} = \Delta E_c + \frac{kT}{q} \ln((m_2/m_1)^{3/2}) \qquad (4.10)$$

Here we have taken that, as follows from figure 4.2, the conduction band offset is equal to $\Delta E_c = \phi_1 - \phi_2$, and also we used the Richardson constant, described by equation (1.8) from chapter 1. Equation (4.10) suggests that the temperature dependence of V_{inf} should be a linear function with the slope depending on the ratio m_2/m_1 between effective masses of carriers in a substrate and LiNbO$_3$ film. The intercept of this linear graph with a vertical axis gives the conduction band offset ΔE_c in a heterojunction. Figure 4.12 demonstrates temperature dependence $V_{\text{inf}}(T)$ of sample LN135.

Analysing temperature dependence $V_{\text{inf}}(T)$, shown in figure 4.12, we obtained the following parameters $\Delta E_c = 0.16$ eV, $m_2/m_1 = 177$ [4]. Furthermore, using the barrier height ϕ_2, obtained earlier, we calculate the Schottky barrier at LiNbO$_3$ site as follows $\phi_1 = \phi_2 + \Delta E_c = 0.43$ eV.

Next, let us analyse the I–V characteristics of the studied samples at high electric fields when the current is limited by the bulk properties of LiNbO$_3$ films. These I–V curves can be described by an equation, similar to (4.6). An original method to determine the most probable charge transport mechanism in the diode-like heterostructures was proposed in work [20]. This approach is based on the analysis of temperature dependence of parameter V_0 in equation (4.6), which is determined by the slope of I–V characteristics in $\ln J$–V coordinates. Figure 4.13 shows temperature dependence of $V_0(T)$, attributed to I–V characteristics of the samples studied and analysed in our work [5].

According to the classification, proposed in [20], curve 1 in figure 4.13 corresponds to the field emission mechanism when parameter V_0 does not depend on temperature. Curve 2 corresponds to the Richardson–Schottky emission, which totally agrees with the results for heterostructures, similar to sample LN134 and described in chapter 3. Curve 3 takes place if the thermally-assisted tunneling

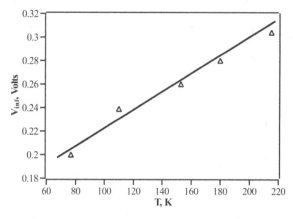

Figure 4.12. Temperature dependence of the voltage, associated with inflection point on I–V curves for sample LN135.

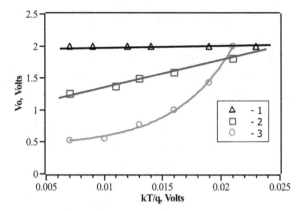

Figure 4.13. Temperature dependence of $V_o(T)$ (V_o versus kT/q plots) for the studied samples at high applied electric fields. 1 sample LN133, 2 sample LN134, 3 sample LN135.

dominates and the intercept of a corresponding horizontal section of the graph with a vertical axis gives the value of V_{oo}, which is defined as (see chapter 1):

$$V_{oo} = \frac{h}{4\pi}\sqrt{\frac{N_d}{m^*\varepsilon\varepsilon_0}} \qquad (4.11)$$

Here N_d is the concentration of ionized donors in a dielectric layer, hn is Planck constant, m^* is the effective mass of the carriers, ε is the dielectric constant of a material.

As was shown in chapter 1 (see equation (1.21), in the case of field emission when non-activated hopping conductivity over CLC (defects) occurs, current density is described in the following way:

$$J = J_0 \exp(-(E_0/E)^{1/4}) \qquad (4.12)$$

Here E is the applied electric field, J_0 is a field independent constant, E_0 is the characteristic field, which is denoted by the following formula:

$$E_0 = \frac{16}{D(E)a^4q} \qquad (4.13)$$

Here $D(E)$ is the energy density of localized states near the Fermi level, q is the electron charge, a is the localization length which is accepted as $a \approx 3a_0$ [21] where a_0 is the crystal lattice parameter.

As follows from equation (4.12), $I–V$ characteristics are the straight lines in $\ln(J)$ versus $E^{-1/4}$ coordinates with a slope giving E_0. Figure 4.14 shows $I–V$ characteristic of sample LN133 at high electric fields in $\ln(J)–E^{-1/4}$ coordinates.

Figure 4.14 clearly demonstrates a linear section in a full agreement with equation (4.12). Taking the lattice parameter for LiNbO$_3$ as $a_0 = 5.1$ Å, we estimated the energy density of states near the Femi level in [5]. On the other hand, the average hopping range and effective concentration of CLC in the framework of this mechanism are described by:

$$R = (q \cdot D(E) \cdot E)^{-1/4} \qquad (4.14)$$

$$N_t = \frac{2}{3\pi R^3} \qquad (4.15)$$

Applying equations (4.14) and (4.15) for sample LN133 we obtain parameters $R = 9.0$ Å, $N_t = 6 \times 10^{17}$ cm^{-3}.

As regards sample LN135, in the framework of the thermally-assisted tunneling, parameter V_{00}, obtained by extrapolation of $V_o(T)$ dependence to $T \rightarrow 0$, gives the donor concentration using equation (4.11) $N_d = 4.0 \times 10^{18}$ cm^{-3} [5]. Moreover, analysis of equation (1.14) suggests that it should be a straight line in $\ln(J_s V_0/T) - 1/V_0$ coordinates with a slope equal to the barrier height φ_b. Since in the present case we deal with bulk-limited conductivity, φ_b can be interpreted as an intergranular barrier. The $\ln(J_s V_0/T) - 1/V_0$ plot for sample LN135 is shown in figure 4.15.

The intergranular barrier, derived from the slope of this graph is $\varphi_b = 0.7$ eV.

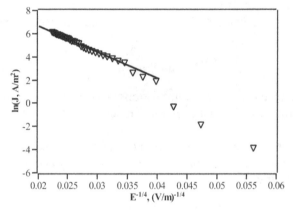

Figure 4.14. Electric field dependence of current density for sample LN133.

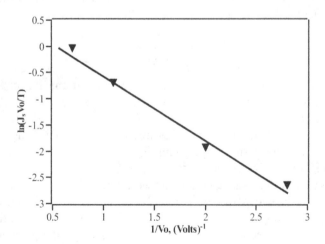

Figure 4.15. The $\ln(J_s V_0/T) - 1/V_0$ plot for sample LN135.

4.1.3 Impedance spectroscopy and ac conductivity of Si–LiNbO$_3$–Al heterostructures

Nyquist diagrams and frequency dependences of the imaginary part of complex impedance Z'' and dielectric modulus M'' for sample LN133 at different temperatures are shown in figure 4.16.

It follows from figure 4.16(a) that we are dealing with conducting films where dielectric effects are minimal. As in chapter 3, we will here analyse the studied sample using the equivalent circuit shown in figure 1.19 (see chapter 1), where R_b and C_b represent resistance and capacitance of LiNbO$_3$ grain's bulk, whereas R_{gb} and C_{gb} are the resistance and capacitance of the grain boundaries. Taking into account the possible distribution of relaxation times, complex impedance of this equivalent circuit can be written as:

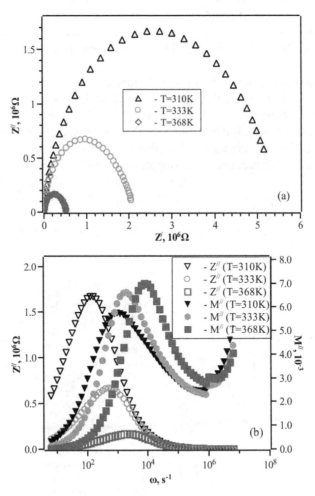

Figure 4.16. Nyquist diagrams (a) and frequency dependences of imaginary part of complex impedance Z'' and dielectric modulus M'' (b) for sample LN133.

$$Z^* = \frac{R_b}{1 + (j\omega\tau_1)^{1-\alpha_1}} + \frac{R_{gb}}{1 + (j\omega\tau_2)^{1-\alpha_2}} \qquad (4.16)$$

Here $\tau_1 = R_b C_b$, $\tau_2 = R_{gb} C_{gb}$ are the relaxation times, j is the imaginary unit and α_i is the coefficient $(0 \leqslant \alpha_i \leqslant 1)$, describing the width of distribution spectrum for relaxation time. Distribution of the relaxation times follows from the fact that Nyquist diagrams are deformed and deviate from ideal semicircles (see figure 4.16). Using both spectra $Z''(\omega)$ and $M''(\omega)$, shown in figure 4.16, and methods of impedance spectroscopy, described in chapter 3, we obtain for sample LN133 the following parameters: $R_{gb} = 3.3 \times 10^6$ Ohms, $C_b = 7.7 \times 10^{-10}$ F [22]. Also, using peak frequencies $\omega_m = (\tau)^{-1} = (RC)^{-1}$, attributed to the same spectra, two other parameters of the equivalent circuit are determined: $C_{gb} = 2.2 \times 10^{-9}$ F, $R_b = 1.3 \times 10^6$ Ohms. Characteristic times, attributed to the response of bulk of grains and grain boundaries in sample LN133 are $\tau_1 = 7 \times 10^{-3}$ s and $\tau_2 = 4 \times 10^{-4}$ s, respectively. By fitting theoretical equation (4.16) to experimental results we obtain coefficients $\alpha_1 = 0.16$ and $\alpha_2 = 0.26$. The experimental and theoretical Nyquist plots, calculated through equation (4.16) with the use of parameters, mentioned above, are presented in figure 4.17.

As was shown in chapter 2 and [1], at higher reactive gas pressure in a reactive chamber (sample LN133), polycrystalline films with arbitrary grain orientation having two-phase composition ($LiNbO_3$ and $LiNb_3O_8$) are formed in the RFMS process. The authors of [23] demonstrated, that under certain synthesis conditions, precipitation of $LiNb_3O_8$ crystallites keeping the epitaxial ratio with initial $LiNbO_3$ crystallites occurs. This is caused by desorption of Li_2O and oxygen atoms out of the inner space of $LiNbO_3$ grains, 'covered' by shells of the $LiNb_3O_8$ phase. In the first approximation, two-phase films can be represented as consisting of grains constructed from a $LiNbO_3$ core and grain boundaries, composed of the $LiNb_3O_8$ phase. Denoting d_b and d_{gb} as the bulk core and grain boundary widths, respectively,

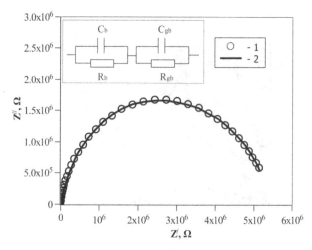

Figure 4.17. Nyquist plots for sample LN133. 1 experiment, 2 theoretical curve, calculated through equation (4.16) with the use of equivalent circuit, shown in the inset.

and using the expression for capacitance of a parallel plate capacitor, enable estimation of the average width of grain boundaries d_{gb} using the following system of equations:

$$\begin{cases} d_g = d_{gb} + d_b \\ \dfrac{d_{gb}}{d_b} = \dfrac{\varepsilon_{gb} C_b}{\varepsilon_b C_{gb}} \end{cases} \qquad (4.17)$$

here d_g is the average grain size. From equation (4.17) the width of grain boundaries can be expressed as:

$$d_{gb} = \frac{d_g}{1 + \varepsilon_b C_{gb}/\varepsilon_{gb} C_b} \qquad (4.18)$$

Using values for C_{gb} and C_b, calculated above and taking into account that $d_g \simeq 50$ nm [1], $\varepsilon_{gb}(LiNb_3O_8) = 34$ [24] and $\varepsilon_b(LiNbO_3) = 28$ [14] we obtain the following results: $d_{gb} = 15$ nm, $d_b = 35$ nm and $d_{gb}/d_b \approx 0.4$ [22].

Grain boundary-limited conductivity is affected by charge transport mechanisms under alternating signals. To investigate ac conductivity, resistances R_b and R_{gb}, derived from impedance spectroscopy, were recalculated in conductivity σ via a simple relationship: $R = d/(\sigma S)$ (here d is the thickness of d_b or d_{gb}, and S is the contact area). Conductivities σ_{gb} and σ_b for sample LN133 at different temperatures are shown in figure 4.18 in Arrhenius coordinates.

As follows from figure 4.18, grain boundary conductivity is an activated process obeying the following law:

$$\sigma_{gb} = \sigma_0 \cdot \exp\left(-\frac{qE_a}{kT}\right) \qquad (4.19)$$

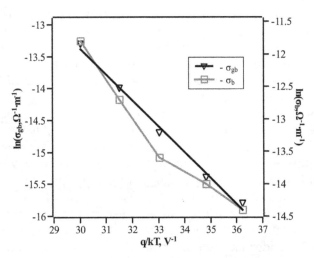

Figure 4.18. Temperature dependence of grain boundaries (σ_{gb}) and bulk of grains (σ_b) for sample LN133 in Arrhenius coordinates.

Here E_a is the activation energy which can be obtained from the slope of a linear $\ln(\sigma) - q/kT$ graph. Activation energy of $E_a = 0.4$ eV for sample LN133 is in good agreement with those, determined for Si–LiNbO$_3$ heterostructures, fabricated at lower working pressure in a reactive chamber [5, 25]. It was demonstrated that at low biases in those heterostructures hopping conductivity dominates when carriers overcome an inter-granular potential barrier of 0.4 eV. The very close value for activation energy was obtained in [26], where conductivity is also limited by the hopping mechanism. Therefore, reactive gas pressure as well as the presence of the LiNb$_3$O$_8$ phase does not influence inter-granular barrier height. Apparently, defects, segregated at the grain boundaries, trap electrons from grains, causing the Fermi level pinning at the interfaces.

With regard to conductivity attributed to a bulk component of polycrystalline grains σ_b, it is affected by two activated processes with the energies of 0.21 eV (at temperatures $T = 300$–350 K) and 0.77 eV (at temperatures $T = 350$–390 K) (see figure 4.18). From our point of view conductivity σ_b is influenced by two mechanisms: correlated barrier hopping conductivity over CLC with a barrier of 0.2 eV (at room temperature) and thermal activation of carriers from deep centers with energy of 0.7 eV in the band gap of LiNbO$_3$ at temperatures $T = 350$–390 K. Activation energy of conductivity $E_a = 0.7$ eV was reported by other authors [27, 28], who attribute it to the trap centers, associated with antisite defects $\text{Nb}_{\text{Li}}^{4\bullet}$ in LiNbO$_3$.

As discussed in chapter 3, in disordered materials ac conductivity obeys the following 'universal' law:

$$\sigma_{ac}(\omega, T) = A(T)\omega^s \qquad (4.20)$$

Here $A(T)$ is the frequency-independent parameter, ω is the angular frequency, s the exponent, which depends on the particular conductivity mechanism, and can be derived from the slope of the graph $\ln\sigma$–$\ln\omega$. This frequency dependence for sample LN133 is shown in figure 4.19.

Three regions, corresponding to different values of the exponent s in equation (4.20) are clearly seen in figure 4.19. The first area at low frequencies corresponds to hopping dc conductivity with activation energy, corresponding to the barrier of $\varphi = 0.4$ eV. Regions 2 and 3 in figure 4.19 correspond to two other charge transport mechanisms. Figure 4.20 illustrates the temperature dependence of exponent s_1 in equation (4.20) for region 2 in figure 4.19.

The diffusion-controlled relaxation model (DCR), describing the charge transport [29], predicts the exponent as $s_1 \approx 0.5$, which is considerably lower than that for the correlated-barrier hopping (CBH) conductivity ($s \approx 0.8$). According to the DCR model charge transport occurs due to diffusive motion between two energetically stable states with activation energy, having two components. First the Coulombic component is attributed to motion of a charge from its oppositely charged surroundings to a position between two adjusted sites. The second component is associated with energy of deformation, involved when a carrier penetrates through a 'sluice', formed by bridge-like atoms, separating two adjusted states. In the framework of this model the exponent s_1 is given:

$$s_1 = 1 - \beta_{CD} \tag{4.21}$$

Here, the Cole–Davison parameter, which in general depends on the ratio τ/τ_d [29], τ and τ_d are the relaxation times, attributed to the diffusion-independent and depended processes, respectively. In a special case when $\omega \gg \tau_d/\tau^2$ we have $\beta_{CD} \rightarrow 0.5$, which is observed in our sample. For a more detailed description of charge transport mechanisms let us analyse the temperature dependence of conductivity in the moderate frequency range ($\omega = 10^3$–10^6 s^{-1}). Figure 4.21 shows the temperature dependence of conductivity for the sample LN133 in the Arrhenius coordinates, corresponding to the second range in figure 4.19.

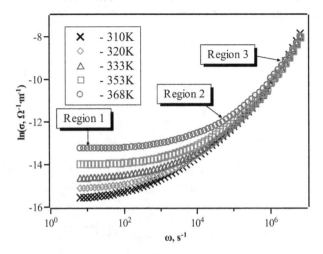

Figure 4.19. Frequency dependence of conductivity for the sample LN133 at various temperatures.

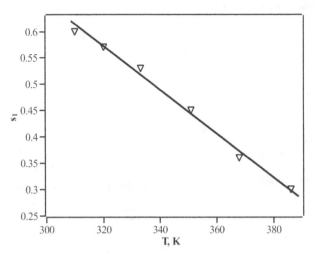

Figure 4.20. Temperature dependence of exponent s in equation (4.20), corresponding to region 2 in figure 4.19 for sample LN133.

As follows from figure 4.21, in this frequency range two activation processes take place with energies of $E_{a1} = 0.03$ eV and $E_{a2} = 0.2$ eV, corresponding to two linear sections at low and high temperatures, respectively. Some authors associate the process, characterized by low activation energy, E_{a1} with electronic hopping conductivity. The process, with energy E_{a2}, is also described in the framework of the DCR model, when activation energy of ac conductivity E_a^{ac} is connected to the dc conductivity E_a^{dc} as $E_a^{ac} \approx (1 - s)E_a^{dc}$ [30] where s is the exponent in equation (4.20). In the studied frequency range this relation is fulfilled because $E_a^{ac} = 0.2$eV, $E_a^{dc} = 0.4$eV, $s = 0.5$.

With regard to the third region in figure 4.19, it was shown in [22], that exponent s_2 in this range has another temperature dependence, shown in figure 4.22. This indicates a new charge transport mechanism.

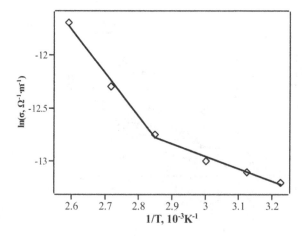

Figure 4.21. Temperature dependence of conductivity in the Arrhenius coordinates for the sample LN133, corresponding to region 2 in figure 4.19.

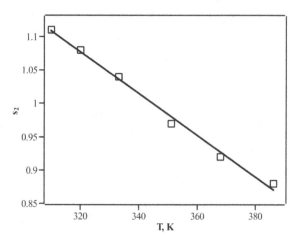

Figure 4.22. Temperature dependence of exponent s_2 in equation (4.20), corresponding to region 3 in figure 4.19 for the sample LN133.

According to the atomic double-well potential model (ADWP model), discussed in [30], conductivity σ_{ac} occurs when carriers overcome exponentially distributed potential barriers. In the framework of the ADWP model, the temperature dependence of exponent s obeys the following law [30]:

$$s_2 = 1 - T/T_o \qquad (4.22)$$

Here T_o is the characteristic temperature in the potential barrier distribution, which can be derived from the slope of a linear graph $s(T)$ (see figure 4.22), which in our case is equal to $T_o = 312$ K and corresponds to an energy scattering of 0.03 eV.

Now let us analyse sample LN134. Unlike sample LN133, in this case dielectric effects rather than conduction phenomena predominate, which is reflected in a diagram similar to the Cole–Cole diagram, shown in figure 4.23.

The $C' - C''$ diagram is more useful than the traditional Cole–Cole diagram, reflecting relationships between imaginary and real parts of the complex dielectric permittivity, because it allows the contribution of various components in capacitance to be analysed.

Two semicircle-like sections (marked by dashed lines), corresponding to low and high frequency ranges are presented in figure 4.23. It is worth stressing that the large semicircle (a low frequency response) is observed only in the temperature range of 95–145 °C and it disappears with a decrease of temperature when only one arc is presented. Now, we analyse the studied heterostructures using an equivalent circuit. However, in contrast to the circuit, shown in the inset in figure 4.17, we use a modified equivalent circuit, demonstrated in figure 4.24.

In the equivalent circuit R_b and C_b are the bulk resistance and capacitance of grains, R_{gb} and C_{gb} are the resistance and capacitance of grain boundaries, R_t and C_t are the resistance and capacitance, caused by the presence of a deep level in the band gap of $LiNbO_3$. Figure 4.25 shows the energy band diagram of two materials (two $LiNbO_3$ polycrystalline grains), containing deep traps and shallow donors with energies E_t and E_d, respectively, as well as interface states.

Among numerous possible processes influencing relaxation phenomena in the studied films, the most often observed ones are marked by the numbers in figure 4.25. At high frequencies, when only charge at the edge of depletion zone, corresponding to coordinate x_d (see figure 4.25) follows the ac signal, measured capacitance is equal to a capacitance of the depletion layer and corresponds to C_{gb} in figure 4.24. At lower frequency and relatively high temperature, when a parameter, corresponding to the thermal emission rate from deep traps is higher than the test frequency ω, differential charge, generated by capture and re-emission of carriers from deep levels contributes capacitance. This change of charge occurs at the point where the Fermi level intersects the deep level E_t (crossover point with coordinate x_t in figure 4.25). This exchange by carriers, marked '1' in figure 4.15, is taken into account by the capacitor C_t and resistor R_t in the equivalent circuit, shown in figure 4.24. Moreover, charge exchange with interface states (process '3' in figure 4.25) also can affect the measured capacitance. Thus, at low frequency the total measured capacitance is influenced by 'slow' processes '1' and '3' in figure 4.25, occurring at the grain boundaries. At high frequency the main

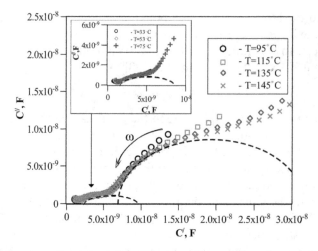

Figure 4.23. Frequency dependence of real C' and imaginary C'' parts of complex capacitance $C^* = C' - jC''$ (Cole–Cole-like diagram) for sample LN134 at different temperatures. The arrow indicates the frequency increase direction.

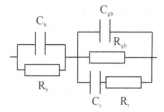

Figure 4.24. Modified equivalent circuit of LiNbO$_3$ films, taking into account the contribution of deep traps to the measured capacitance.

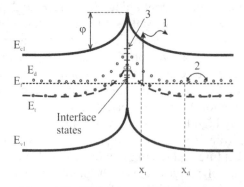

Figure 4.25. Band diagram of LiNbO$_3$ grains.

contribution is attributed to the bulk of grains where the 'fast' process of hopping conduction via shallow CLC in band gap of LiNbO$_3$ dominates (see figure 4.25). In the framework of this model each semicircle in figure 4.23 corresponds to one of the parallel elements in the equivalent circuit, shown in figure 4.24. The complex capacitance, corresponding to this equivalent circuit can be expressed as:

$$C^* = C_{gb} + \frac{C_t}{1 + j\omega\tau} - \frac{j}{\omega R_{gb}} \qquad (4.23)$$

Here $\tau = R_t C_t$ is the relaxation time corresponding to the response of a deep level with energy of E_t in the band gap of LiNbO$_3$. The total capacitance of a depletion layer is determined by the deep level when the test signal frequency is [31]:

$$\omega_t = 2e_n'\left[2 - \frac{w - x_t}{w}\right] \qquad (4.24)$$

Here w is the width of the depletion layer, e_n' is the the thermal excitation coefficient from deep levels, which is described by the following expression:

$$e_n' = C_n N_c \exp\left(-\frac{E_c - E_t}{kT}\right) \qquad (4.25)$$

where C_n is the capture coefficient of a trap level. At the frequency of $\omega = \omega_t$ a maximum should occur on both the frequency dependence of capacitance $C''(\omega)$ and the dielectric loss tangent tg $\delta(\omega)$. It can be shown that the real and imaginary parts of complex capacitance in equation (4.23) are described by the following expressions:

$$C' = \frac{C_t}{1 + (\omega\tau)^2} + C_{gb}$$
$$C'' = \frac{C_t \omega\tau}{1 + (\omega\tau)^2} \qquad (4.26)$$

Capacitance, associated with a deep level response can be found as [32]:

$$C_t = \frac{q^2}{kT} N_t f_0 (1 - f_0) S \qquad (4.27)$$

Here N_t is the density of CLC, f_0 is the the Fermi function and S the contact area.

The imaginary part of complex capacitance C'', described by equation (4.26) goes through a maximum when $\omega\tau = 1$, which gives a relaxation time of $\tau = 1/\omega$. The second equation (4.26) at the maximum is:

$$C''|_{max} = \frac{C_t}{2} \qquad (4.28)$$

Using equation (4.27) for C_t, we obtain the density of CLC N_t through equation (4.28) in the following form:

$$N_t = \frac{8kT}{q^2 S} C''|_{max} \qquad (4.29)$$

The energy of deep traps E_t can be found from the slope of a linear graph $\ln(\omega_t)$ − q/kT (see (4.24) and (4.25)). Figure 4.26 demonstrates the temperature dependence of the dielectric loss tangent for the sample LN134 with two peaks, clearly observed at low frequencies (peak 1) and high frequencies (peak 2).

Each peak in figure 4.26 is attributed to the 'slow' and 'fast' relaxation processes, occurring at the grain boundaries and in the bulk of grains, respectively. Temperature dependence of a frequency ω_m at the maximum of spectra $tg(\delta) - \omega$ for both peaks is shown in figure 4.27.

It can be seen from figure 4.27, that temperature dependences of ω_m for both peaks are linear functions $\ln(\omega_m) - q/kT$ with the slopes, giving activation energies $E_{a1} = 0.9$ eV and $E_{a2} = 0.5$ eV for peak 1 and peak 2, respectively, in full agreement with equation (4.25). Determining $C''|_{max}$ from the Nyquist diagrams and using

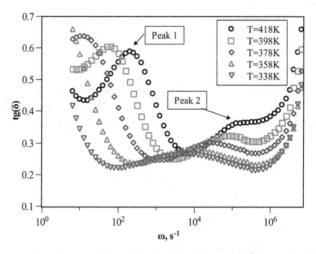

Figure 4.26. Frequency dependence of the dielectric loss tangent for the sample LN134 at different temperatures.

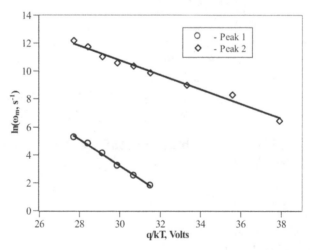

Figure 4.27. Temperature dependence of maximum frequency ω_m, corresponding to the two peaks in figure 4.26.

equation (4.29) we estimated the density of states for traps with energy $E_{t1} = 0.9$ eV in sample LN134 as $N_t = 7.8 \times 10^{17}$ cm^{-31}.

Some researchers demonstrate [33, 34] that activation energy of 0.9 eV in LiNbO$_3$ corresponds to an energy level E_t, associated with the presence of lithium vacancies. On the other hand, activation energy of 0.5 eV is attributed to the hopping polaron conductivity, dominating in LiNbO$_3$ single crystals according to many works [35–38]. It was shown in earlier works, that in the case of oxygen deficit the charged oxygen vacancies and free electrons occur in nonstoichiometric LiNbO$_3$ single crystals [39, 40]. In contrast, some authors state, that oxygen vacancies are not major defects, because in this case oxygen and lithium diffuse, leaving antisite Nb$_{Li}^{4\bullet}$ and lithium vacancies, which is reflected in the increase in density of LiNbO$_3$ crystals [9]. However, recent studies demonstrate, that thermal annealing of single crystals in pure oxygen leads to a decrease in dc conductivity, caused by the neutralization of oxygen vacancies [35]. The effect of the oxygen deficit on dc conductivity can be described by the following equation [41]:

$$2Nb^{5+} + O^{2-} \Leftrightarrow 2Nb^{4+} + O_v^{2\bullet} + \frac{1}{2}O_2 \qquad (4.30)$$

Here $O_v^{2\bullet}$ is a doubly-charged oxygen vacancy. Thus, free electrons are generated due to ionization of Nb^{4+} ions according to the following reaction:

$$Nb^{4+} \Leftrightarrow Nb^{5+} + e^- \qquad (4.31)$$

Consequently, electrons, that have escaped from the neutral oxygen, become free or self-trapped in the positions of Nb ions, forming small polarons and influencing electrical conductivity. This point of view agrees with the n-type conductivity of LiNbO$_3$ single crystals with oxygen reduction [35]. It was clearly demonstrated in [36] that activation energy of $W_h = 0.35$–0.4 eV corresponds to hopping polaronic conductivity when polaron binding energy lies in the range of $W_p = 0.7$–0.8 eV. Thus, when in our experiments the thermally activated conductivity with activation energy of 0.7 eV is observed it can be attributed to thermal activations of polarons, associated with antisite defects Nb$_{Li}^{4\bullet}$ [41]. Furthermore, some researchers reported hopping dc conductivity of small polarons with energy of $E_a = 0.05$ eV (corresponding to the polaronic band in lithium niobate [36]), which is in good agreement with our results for samples similar to LN134 studied in chapter 3.

Nyquist diagrams for the sample LN135 at three temperatures regimes are shown in figure 4.28(a).

It follows from the Nyquist plot, that the response of the studied heterostructures on ac signal results from their conductivity rather than dielectric phenomena. As we have discussed, the studied sample can be reasonably represented by two Schottky barriers connected back-to-back in series (see figure 4.28(b)). Thus, for further analysis of impedance spectra we can use a simplified equivalent circuit, shown in

---

[1] This value was re-calculated from the surface density into bulk concentration.

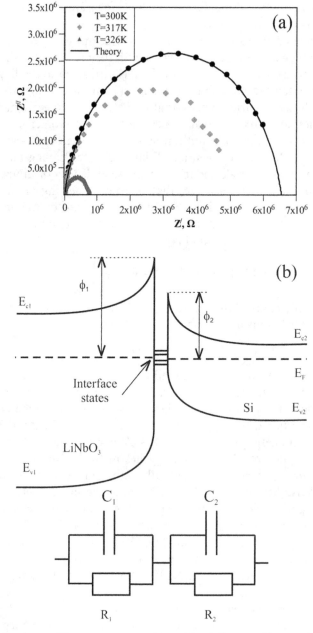

Figure 4.28. Nyquist plots at different temperatures (a), and band diagram with an equivalent circuit (b) for sample LN135.

figure 4.28(b). Here C_1 and C_2 represent the capacitances of depletion zones in LiNbO$_3$ and Si, respectively, whereas the resistors R_1 and R_2 correspond to their resistances. The complex impedance for such equivalent circuit is described by an equation similar to (4.16). Both parallel RC elements in the equivalent circuit should be represented by two semicircles in appropriate frequency ranges in the Nyquist diagrams. A large semicircle corresponds to a component with the largest resistance in the non-uniform system. We have demonstrated that conductivity of the sample LN135 is described by the double depletion layer model and it is affected by the Schottky barrier heights ϕ_1 and ϕ_2 (see figure 4.28(b)). It follows from I–V analysis that $\phi_1 > \phi_2$, so the highest resistance (and the largest semicircle in the Nyquist diagram) corresponds to the depletion zone in Si. Carriers at this temperature have sufficient energy to only overcome the lowest barrier. At room temperatures, the larger semicircle masks the smaller one, corresponding to the

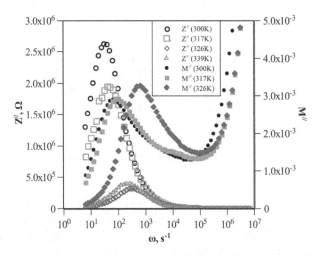

Figure 4.29. Frequency dependences of imaginary parts of complex impedance Z^* and dielectric modulus M^* for sample LN135 at different temperatures.

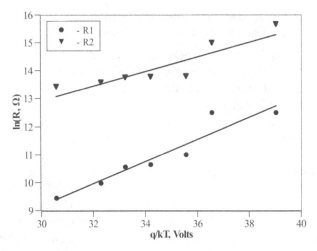

Figure 4.30. Temperature dependence of resistances R_1 and R_2 in the equivalent circuit, shown in figure 4.28(b).

response of the depletion zone in LiNbO$_3$, which starts dominating when temperature increases (see figure 4.28(a)). To distinguish the contributions from both parallel RC elements in total impedance we have measured the frequency dependences of imaginary parts of complex impedance Z^* and dielectric modulus M^*, shown in figure 4.29.

Both spectra clearly demonstrate peaks, corresponding to maximal resistance (a peak on $Z''(\omega)$) and minimal capacitance (a peak on $M''(\omega)$). Based on the approach described earlier, we have determined the values of all elements in the equivalent circuit. Note, that coefficients α_1 and α_2 describing distribution in relaxation times in equation (4.16), are close to zero, which indicates nearly single characteristic times with values of $\tau_1 = 8 \times 10^{-3}$ s and $\tau_2 = 5 \times 10^{-2}$ s. Temperature dependences of resistances R_1 and R_2 in Arrhenius coordinates are shown in figure 4.30.

Activation energies, determined from the slope of corresponding straight lines in figure 4.30 have magnitudes of 0.4 eV and 0.26 eV, which correspond to the Schottky barriers of ϕ_1 and ϕ_2, derived from I–V analysis.

Table 4.4 summarizes the data associated with the influence of sputtering parameters on electrical properties of studied LiNbO$_3$-based heterostructures.

4.2 Thermal annealing effect on electrical properties of Si–LiNbO$_3$ heterosystem

It was demonstrated in chapter 2 that thermal annealing (TA) positively influences the degree of crystallinity, surface roughness of synthesized films and reduces mechanical stress in them. Thus, TA should affect electrical properties of Si–LiNbO$_3$ heterostructures as well. Similar to the study of structural properties, TA of the studied films was performed at temperatures of 600–700 °C in a coaxial in an oxygen atmosphere oven for one hour.

4.2.1 Capacitance–voltage characteristics and ferroelectric properties of Si–LiNbO$_3$–Al heterostructures after thermal annealing

Figure 4.31 shows P–E loops of LiNbO$_3$ films after TA. All parameters of P–E loops shown in figure 4.31 are given in table 4.5. It follows from table 4.5 that TA does not influence remnant polarization despite the presence of a not ferroelectric LiNb$_3$O$_8$ phase in all films after TA (see chapter 2). On the other hand, TO leads to shifting all P–E loops along the vertical axis regardless of the sputtering regimes. Such behavior can be explained by the preferable orientation of polarization in the films after TA.

In fact, the local domain structure of synthesized films changes significantly in the process of TA. Figure 4.32 demonstrates the local domain structure of as-grown LiNbO$_3$ films and films after TA and obtained by the PFM method.

Three contrasts are observed in the vertical piezoresponse component (figure 4.3) light, dark and intermediate. The light and dark components correspond to ferroelectric domains with preferable vertical direction (up or down), whereas the intermediate component corresponds to polarization along the surface of a sample.

Table 4.4. Summarized data on the influence of reactive gas composition and pressure on electrical properties of Si–LiNbO₃–Al heterostructures.

Properties	Sample #		
	Increase of pressure. The presence of oxygen in a chamber →		
	LN133	LN134	LN135
Effective density of charge in a film, Q_{ef} (C cm⁻²) (possible sources: charged antisite defects $Nb_{Li}^{4\bullet}$ and oxygen vacancies)	$+9.0 \times 10^{-7}$	$+8.0 \times 10^{-7}$	$+5.1 \times 10^{-8}$
Effective density of CLC (defects), N_{ef}, cm⁻²	5.7×10^{12}	5.0×10^{12}	3.2×10^{11}
Donor concentration in Si, N_d (cm⁻³) (nominal $N_{do} = 1 \times 10^{15}$ cm⁻³). Donor formation is cause by diffusion of oxygen and lithium	Declines exponentially from 2×10^{16} at the interface to N_{do} at the distance of 550 nm.	Declines exponentially from 2×10^{16} at the interface to N_{do} at the distance of 550 nm.	Declines exponentially from 2×10^{18} at the interface to N_{do}.
Donor concentration in LiNbO₃ film, N_d (cm⁻³)		7.0×10^{13} (I–V analysis)	7×10^{17} (C–V analysis)
Density of states at the Si/LiNbO₃ interface, D_{ss} (eV⁻¹ cm⁻²)	5.0×10^{12}	9.0×10^{12}	2.5×10^{13} (caused by the formation of an intermediate layer at the Si/LiNbO₃ interface).
Band diagram	Figure 4.9	Figure 3.25(b) (see chapter 3)	Figure 4.28(b)
DC conductivity	Low electric fields (<3.3 kV cm⁻¹): Hopping conductivity limited by intergranular potential barrier of 0.4 eV in LiNbO₃ films. Hopping conductivity over CLC with activation energy of 0.2 eV. At high temperatures is caused by thermal ionization of carriers from deep levels ($E_t = 0.7$ eV).	Low electric fields (<5 kV cm⁻¹) hopping polaronic conductivity with activation energy of 0.05 eV over CLC.	Average electric fields (3.3–10 kV cm⁻¹): the Richardson–Schottky emission over the Schottky barrier of $\phi_1 = 0.43$ eV at Si/LiNbO₃ interface.

(Continued)

	Average fields (3.3–10 kV cm⁻¹): at temperatures of $T = 77$–200 K: the Fowler–Nordheim tunneling through the potential barrier of $\phi = 0.4$ eV at Si/LiNbO₃ interface. At $T \approx 300$ K: the Richardson–Schottky emission. Strong fields (>10 kV cm⁻¹): at $T \approx 300$ K: non-activated hopping conductivity over CLC in LiNbO₃ films.	Average fields (5–30 kV cm⁻¹): the Fowler–Nordheim tunneling through the potential barrier of $\varphi_b = 0.25$ eV at Si/LiNbO₃ interface. Strong fields (30–90 kV/cm): the Richardson–Schottky emission over the barrier of 0.02 eV.	Strong fields (>10 kV cm⁻¹): the thermally-assisted tunneling through the intergranular potential barrier of 0.7 eV in LiNbO₃ films.
AC conductivity and dielectric losses	Low frequency ($\omega = 10$–10³ s⁻¹) response: from grain boundaries (apparently by LiNb₃O₈ phase) with average relaxation time of $\tau_1 = 1 \times 10^{-3}$ s and with correlated-barrier hopping conductivity. Average frequency ($\omega = 10^3$–10⁵ s⁻¹) response: hopping conductivity over CLC (the diffusion-controlled relaxation mechanism) with activation energies of 0.03 eV ($T = 300$–350 K) and 0.2 eV ($T = 350$–390 K).	Dielectric effects dominate over conductivity. Low frequency ($\omega = 10$–10³ s⁻¹) response: from grain boundaries is attributed to the trapping and re-emission of electrons from the deep traps with relaxation time of $\tau_1 = 130$ s. High frequency ($\omega = 10^3$–10⁵ s⁻¹) response: from bulk of grains (LiNbO₃ phase) with relaxation time of $\tau_2 = 1.5 \times 10^{-3}$ s. Hopping conductivity of small polarons with activation energy of 0.5 eV.	Hopping conductivity limited by the Schottky barriers at LiNbO₃/Si interface with heights of $\phi_1 = 0.43$ eV and $\phi_2 = 0.26$ eV (see the band diagram).

High frequency ($\omega = 10^5$–10^7 s^{-1}) response: from bulk of grains (LiNbO$_3$ phase) with average relaxation time of $\tau_2 = 4 \times 10^{-4}$ s. Hopping conductivity over the exponentially distributed potential barriers.

Concentration of CLC (traps), N_t (cm^{-3})	6×10^{17} (I–V analysis)	7.8×10^{17}	4.0×10^{18} (C–V analysis)
Energy of traps in the bang gap of LiNbO$_3$ films	$E_{t1} = 0.2$ eV, $E_{t2} = 0.7$ eV (attributed to antisite defects $Nb_{Li}^{4\bullet}$)	$E_{t1} = 0.9$ eV (attributed to lithium vacancies V_{Li}'), $E_{t2} = 0.5$ eV	$E_t = 0.7$ eV (attributed to antisite defects $Nb_{Li}^{4\bullet}$)

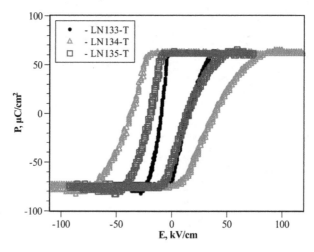

Figure 4.31. Ferroelectric hysteresis loops for LiNbO$_3$ films after TA.

Table 4.5. Parameters of experimental P–E loops for LiNbO$_3$ films after TA (the corresponding parameters of as-grown films are given in brackets).

Sample #	Remnant polarization, P_r (μC cm^{-2})	Coercive field, E_c (kV cm^{-1})	Shift along the vertical axis, ΔP_r (μC cm^{-2})	Build-in field, E_b (kV cm^{-1})
LN133-T	68 (69)	12 (14.0)	−6 (−1)	2.4 (2.8)
LN134-T	69 (69)	37.8 (39.5)	−7 (−1)	1.2 (2.8)
LN135-T	69 (69)	17.4 (27.3)	−7 (−1)	−2.7 (2.0)

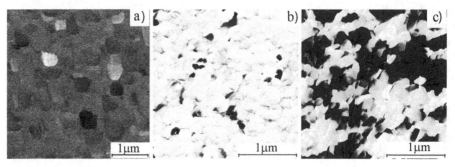

Figure 4.32. PFM images of as-grown LiNbO$_3$ film (a) and films after TA (b),(c). Patterns (a) and (b) correspond to the vertical polarization component, pattern (c) is the lateral polarization component [42]. Reprinted by permission from Springer Nature. Copyright 2013.

As-grown LiNbO$_3$ films do not manifest preferable domain orientation (figure 4.32(a)). The signal of piezoresponse from the films after TA demonstrates one vertical direction of polarization with few inclusions of the opposite direction, whereas laterally, the complex domain structure is observed (figures 4.32(b) and (c)). This result explains the vertical shift of *P–E* loops of the films after TA, in which preferentially oriented domains are observed, that makes them non-symmetric in terms of polarization reversal relative to the applied electric field. Similar behavior of the local domain structure was observed in [43] in which the influence of annealing temperature on ferroelectric properties of LiNbO$_3$ is studied. Annealing temperature of 700 °C is the optimal parameter in terms of preserving local domain structure and the optimal LiNbO$_3$/LiNb$_3$O$_8$ phase ratio.

Our results, given in table 4.5 reveal the other interesting properties of the studied films after TA. The coercive field of the films, fabricated by RFMS in an Ar(60%) + O$_2$(40%) gas mixture, decreases dramatically after TA (sample LN135-T). Furthermore, TA of these films leads to formation of relatively high built-in field, with opposite direction compared to as-grown films. To investigate the possible nature of this phenomenon we use analysis of high frequency *C–V* characteristics.

It was shown earlier, that the films, fabricated in an Ar atmosphere, contain a positive oxide charge which does not depend on reactive gas pressure in a chamber, degree of orientation in films and the presence of the LiNb$_3$O$_8$ phase (sample LN133 and LN134). Therefore, we compare *C–V* characteristics of as-grown heterostructures, fabricated in a pure Ar environment and in an Ar + O$_2$ gas mixture (samples LN134 and LN135) with the same heterostructures after TA (samples LN134-T and LN135-T). *C–V* curves of the studied samples are shown in figure 4.33.

TA leads to a decrease of positive oxide charge in the films for both samples, which is reflected in figure 4.33 as the decrease in the horizontal shift of experimental *C–V* curves compared to the ideal *C–V* ones. Furthermore, it follows from figure 4.33(b), TA leads to the disappearance of a branch, attributed to modulation of the depletion zone in LiNbO$_3$ and observed for sample LN135. Under the regimes, corresponding to the sample LN135 an intermediate layer with variable composition is formed between a Si substrate and LiNbO$_3$ films. The absence of a depletion zone modulation in LiNbO$_3$ films in sample LN135-T results in the formation of an abrupt film–SiO$_2$ layer interface in the studied heterostructures after TA.

Also, it is important to stress that *C–V* curves of all samples after TA manifest a hysteresis, shown in figure 4.34.

One possible explanation of this phenomenon could be a migration of positively charged ions in a film, which leads to a charge accumulation during measurements because ions do not follow the change of voltage. However, this hypothesis is not realistic in our case, because *C–V* hysteresis is observed at the liquid nitrogen temperature when ions are 'frozen' and do not take part in the charge transport. On the other hand this counterclockwise hysteresis type can be caused by the capture of positive charge by traps. The magnitude of this charge can be estimated through the width of a 'window' ΔV on the hysteresis curves, using the following expression:

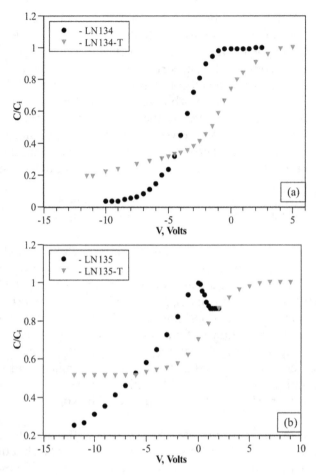

Figure 4.33. Typical high frequency C–V characteristics of as-grown heterostructures (samples LN134 and LN135) and heterostructures after TA (samples LN134-T and LN135-T), recorded at $T = 293$ K.

$$Q_{ot} = C_{FB} \cdot \Delta V_{FB} \qquad (4.32)$$

here C_{FB} is the flat band capacitance, ΔV_{FB} *is* the difference in the flat band voltage of a heterostructure for forward and reverse passage through voltage. We will discuss the origin of this C–V hysteresis later. The results of C–V analysis for heterostructures after TA are given in table 4.6 and in figures 4.35 and 4.36.

As follows from figures 4.36 and 4.6(b) and (c), TA leads to a significant decrease in the energy distribution of surface states at the Si/film heterojunction. Apparently it results in decreasing defects concentration in the SiO_2 layer, existing at the heterojunction. In addition, the donor distribution in a silicon substrate also changes (see figures 4.35 and 4.4). Concentration of donors in sample LN134-T rises with TA, whereas in sample LN135-T TA leads to decreased N_d. This can be explained under the assumption that oxygen, diffusing in silicon, creates deep donors. In the annealing process of Si–LiNbO$_3$ heterostructures, fabricated in an Ar atmosphere

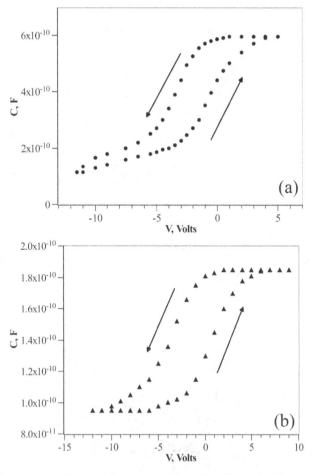

Figure 4.34. High frequency $C-V$ hysteresis loops for sample LN134-T (a) and LN135-T (b) at a temperature of $T = 293$ K.

Table 4.6. Results of $C-V$ analysis for samples LN134-T and LN135-T (corresponding parameters for as-grown heterostructures are given in brackets).

Sample #	Dielectric constant of a film, ε	Flat band voltage, V_{FB} (V)	Effective charge in a film, Q_{eff} (C cm^{-2})	Effective density of states, N_{eff} (cm^{-2})	Position of a charge centroid (relative to a film surface) in thickness units, d_c/d	Trapped charge, Q_{ot} (C cm^{-2})
LN134-T	25	−0.1 (−5.6)	+5.5 × 10^{-9} (+8.0 × 10^{-7})	3.3 × 10^{10} (5.0 × 10^{12})	0.97 (0.36)	1.7 × 10^{-7}
LN135-T	12	2.0 (−3.0)	−4.0 × 10^{-8} (+5.1 × 10^{-8})	2.4 × 10^{11} (3.2 × 10^{11})	0.71	8.5 × 10^{-8}

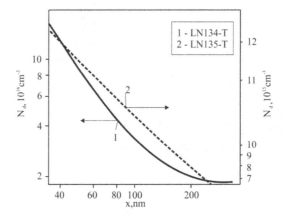

Figure 4.35. Doping profile of the Si–LiNbO$_3$ heterostructures after TA; x is the distance from Si/LiNbO$_3$ interface towards the bulk.

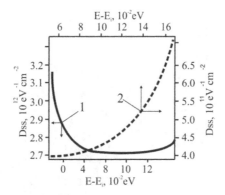

Figure 4.36. Energy distribution of surface states in the upper half of the band gap of Si for the studied Si–LiNbO$_3$ heterostructures after TA (1—sample LN134-T, 2—sample LN135-T).

(sample LN134-T), oxygen actively diffuses in Si under a gradient of concentration, creating additional donor centers. In the case of heterostructures, synthesized in an Ar(60%) + O$_2$(40%) gas mixture (sample LN135-T), the thick SiO$_2$ layer at the Si/film interface limits the oxygen diffusion. On the other hand, the reverse diffusion of O$_2$ (see chapter 2) from Si also reinforces this process.

Evidently, oxygen diffusion is responsible for the decrease in positive oxide charge, formed in as-grown LiNbO$_3$ films (see table 4.6). Molecular oxygen, diffusing during TA, neutralizes positively charged oxygen vacancies in LiNbO$_3$ films as proposed in work [41]. The flat band regime is being implemented in Si–LiNbO$_3$ heterostructures, formed in an Ar environment, after their TA, which is important for application to LiNbO$_3$-based heterostructures when space charge effects in a substrate play a negative role. This result is especially important for practical applications of LiNbO$_3$-based heterostructures in optoelectronics and non-volatile memory units.

It is worth noting, that TA changes the sign of oxide charge from positive to negative in heterostructures, fabricated in an Ar(60%) + O_2(40%) reactive gas mixture (see table 4.6, sample LN135-T). This phenomenon can be explained in the framework of the formation of a complex $V_{Li}' - 1/2O_2$. If molecular oxygen, diffusing into LiNbO$_3$ film, settled near a vacancy, charge can be transferred from a vacancy to oxygen according to the following reaction [44]:

$$V_{Li}' + 1/2O_2 \Rightarrow V_{Li}^0 + 1/2O_2' \tag{4.33}$$

here V_{Li}' and V_{Li}^0 are negatively charged and neutral vacancies, respectively. The decrease of built-in field in LiNbO$_3$-based heterostructures after TA (see table 4.5) also can be explained in terms of the reduction of positive oxide charge in films during TA.

C–V hysteresis shown in figure 3.34 can be attributed to the charge trapping effects in LiNbO$_3$-based heterostructures after TA (see table 4.6). The magnitude of trapped charge does not differ significantly in heterostructures fabricated in different reactive gas environments. According to the classification of oxide charge given in chapter 3, capture and emission of carriers is possible only for the interface-trapped charge (Q_{it}) and the oxide-trapped charge (Q_{ot}). In turn, Q_{ot} can be captured by states, existing at the Si/oxide interface ('interface traps') or distributed near the interface ('border traps') at a distance of about 5 nm from the semiconductor surface [45]. Our structural study (see chapter 2) demonstrates that the SiO$_2$ layer with thickness of 5 nm is formed at the Si/LiNbO$_3$ interface in heterostructures, formed in an Ar atmosphere by RFMS. In Si–LiNbO$_3$ heterostructures fabricated in an Ar (60%) + O_2(40%) gas mixture the extended intermediate layer with composition close to SiO$_2$ is formed. Apparently, thermal annealing forces oxygen to diffuse toward the Si/LiNbO$_3$ interface along with re-crystallization of deposited films. Oxygen, interacting with silicon, leads to the formation of the extended intermediate layer, which is confirmed by the study of elemental composition (see chapter 2). The formation of this layer during TA can be a source of additional CLC, created not only due to specific technological regimes, but also by applied electric fields during electrical measurements. It is well known, that classification of defects depends on their location, electrical behavior, charge state and physical structure. Primarily, the term 'trap' refers to the defect location: 'oxide traps' and 'interface traps' are defects, located in the layer and at the substrate/layer interface, respectively. In contrast, the term 'state' is associated with electrical behavior and we can distinguish between 'fixed' (defects whose charge state does not change) and 'switching' (defects which exchange charge with a substrate or an electrode. The 'fixed' states can be positive, negative or neutral, whereas the 'switching' states change their charge state due to the capture and re-emission of an electron or a hole. They can be donor-like (acceptor-like) when the states are neutral initially and become positively charged due to emission of an electron (capture of a hole), or when they become negatively charged due to the capture of an electron (emission of a hole). Finally, defects can be stable or unstable depending on experimental conditions.

Defects in SiO_2 can be intrinsic and caused by the growth process (for instance oxygen vacancies), or they can be generated by irradiation or applied electric fields [46]. Numerous papers suggest, that avalanche electron or hole injection from the Si substrate and the Fowler–Nordheim electron injection from the substrate are the major sources of defect generation in silicon dioxide in the Si–SiO_2 system. The kinetics of defect generation is quite complex due to varieties of their types: from positively and negatively charged fixed traps to 'border' and 'interface' states. Electrons, injected from the Si substrate through the Fowler–Nordheim tunneling, gain their energy due to high electric fields in the SiO_2 layer (electron heating) [47]. Hot electrons generate electron–hole pairs, provided the thickness of the SiO_2 layer is higher than 10–12 nm at fields >10 MV cm^{-1}. Generated holes tunnel into the SiO_2 layer due to the strong electric field. Some of them can be trapped by traps in the bulk of the layer or at the interface, leading to increase in the positive captured charge. Recombination of electrons with these captured holes can be the major source of neutral electron traps, border traps and interface states at the Si–SiO_2 interfaces [48]. The authors of [49, 50] revealed an important property: the Fowler–Nordheim emission does not generate minority carriers described above, in thin SiO_2 films (<20 nm), because this thickness is not sufficient for heating of tunneling electrons. Evidently, in our case, the thin (~5 nm) SiO_2 layer at Si/$LiNbO_3$ interface in as-grown heterostructures (see chapter 2) is not sufficiently thick for the generation of electron–hole pairs, despite that Fowler–Nordheim emissions are frequently observed in the studied heterostructures [5, 51]. Therefore, C–V curves of as-grown Si–$LiNbO_3$ heterostructures do not manifest the positive charge capture at the border traps and consequently hysteresis does not appear. TA leads to the formation of an intermediate layer with a higher quality of Si/SiO_2 interface, making possible the above described process of generation and capture of holes. To verify this assumption we have studied I–V characteristics of Si–$LiNbO_3$ after TA.

4.2.2 Current–voltage characteristics of Si–$LiNbO_3$–Al heterostructures after thermal annealing

Figure 4.37 demonstrates I–V characteristics of as-grown Si–$LiNbO_3$–Al hetero-structures, formed by rhw RFMS method in an Ar reactive gas environment (sample LN134) and the same heterostructure after TA (sample LN134-T).

Two sections are clearly seen on I–V curves, separated by the dashed line. The first section in the range of relatively low electric fields ($E < 2 \times 10^6$ V m^{-1}) of fast growing current is associated with barrier properties of the Si–$LiNbO_3$ heterostruc-ture. The second, more inclined section, at $E > 2 \times 10^6$ V m^{-1} is attributed to the currents, limited by the film bulk properties. As seen from figure 4.37, TA leads to a decrease in conductivity of as-grown films fabricated in an Ar environment. Furthermore, the linear sections in I–V characteristics plotted in the Fowler–Nordheim coordinates (see insert in figure 4.37), indicate that tunneling dominates in this applied electric fields range. Using equation (4.5) we determine the height of the potential barrier, corresponding to traps with energy of 1.7 eV in the band gap of $LiNbO_3$. This magnitude is almost exactly equal to the energy, determined in chapter 3 and our work [52] for as-grown Si–LiNbO₃ heterostructures. The observed Fowler–Nordheim emission supports the model, discussed above and associated with the generation of holes in the SiO_2 layer and their subsequent capture by border traps. This trapping and re-emission is responsible for $C–V$ hysteresis shown in figure 4.34.

At strong electric fields ($E > 2 \times 10^6$ V m⁻¹) the Ohmic conductivity is observed in a wide temperature range, which follows from a linear $I–V$ dependence in the $\ln(J)$ – $\ln(E)$ coordinates with the slope of 1 (see figure 4.38).

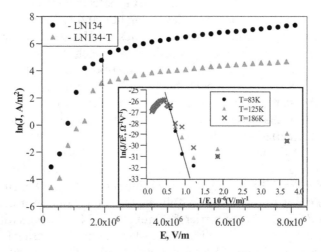

Figure 4.37. $I–V$ characteristics of as-grown Si–LiNbO₃–Al heterostructures, fabricated by RFMS method in an Ar environment (sample LN134) and heterostructures after TA (sample LN134-T) recorded at room temperature. The inset shows $I–V$ characteristics of sample LN134-T in the Fowler–Nordheim coordinates at different temperatures.

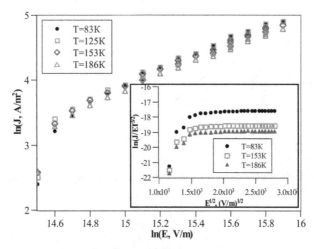

Figure 4.38. $I–V$ characteristics for sample LN134-T in the $\ln(J)$ – $\ln(E)$ coordinates at different temperatures. The inset shows $I–V$ curves in the Simmons coordinates (explanations are in the text).

The most frequently observed Richardson–Schottky emission is described by equation (1.8) in chapter 1, which indicates that I–V characteristics should be linear in the Simmons coordinates: $\ln(J/ET^{3/2}) - \ln(\sqrt{E})$. However, if conductivity is limited by the bulk of a material, the pre-exponential term dominates in equation (1.8):

$$J_0 = 2q\left(\frac{2\pi mkT}{h^2}\right)^{3/2} \mu E_0 \qquad (4.34)$$

Here q is the elementary charge, m is the carrier effective mass, k the Boltzmann's constant, T the temperature, μ the carrier's mobility, and E_0 the applied electric field. Thus, if the pre-exponential factor dominates, I–V characteristic should be a horizontal line in the Simmons coordinates, which is observed for sample LN134-T (see the inset in figure 4.38). Furthermore, as follows from equation (4.34), the intercept of the horizontal branch of I–V characteristic with the vertical axis is proportional to carrier mobility in a film. Temperature dependence of this parameter is determined only by the temperature dependence of the mobility depending on the scattering mechanisms. Figure 4.39 shows temperature dependence of drift mobility in the $\ln(\mu) - \ln(T)$ coordinates, calculated from equation (4.34).

The magnitude of mobility at room temperature is $\mu = 1.3 \times 10^{-12}\ \mathrm{m^2\,V^{-1}\,s^{-1}}$, which is close to the drift mobility in single crystal lithium niobate ($\mu = 8 \times 10^{-13}\ \mathrm{m^2\,V^{-1}\,s^{-1}}$) [53]. It is important to note, that as follows from the slope of a graph, shown in figure 4.39, the drift mobility of carriers in sample LN134-T declines with temperature according to the $T^{-3/2}$ law, which is attributed to phonon scattering [54].

Thus, taking into account the results of structural analysis, we can conclude, that TA of $LiNbO_3$ films leads to their recrystallization, increasing the size of grains and minimizing the influence of intergranular barriers on charge transport. As a result, electrical conductivity of $LiNbO_3$ films results in hopping conductivity with the phonon scattering of electrons as in single crystal lithium niobate.

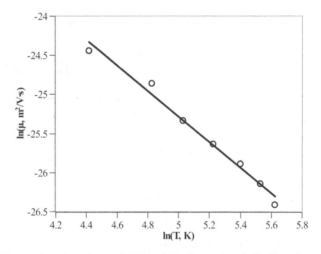

Figure 4.39. Temperature dependence of drift mobility for carriers in the film of sample LN134-T.

As was demonstrated earlier by C–V analysis, TA of Si–LiNbO$_3$ heterostructures, fabricated in an Ar + O$_2$ gas mixture, leads to the lower depletion zone at the substrate/film interface. Thus, in these heterostructures we can expect the absence of a blocking contact at the Si–films interface, which improves its injection properties being a required condition for space charge limited currents (SCLC). Indeed, I–V characteristics of sample LN135-T and other LiNbO$_3$-based heterostructures [55, 56] are linear in the double logarithmic coordinates and have several sections, attributed to SCLC (see figure 4.40).

The power dependence of I–V characteristics is the attribute of SCLC in the form of $J \propto V^\alpha$ where coefficient α (differential slope of I–V curves) depends on concentration and distribution of traps. In many cases when monoenergetic traps are present in the band gap of a material, the linear section with $\alpha = 2$ is observed on the graph $\ln(J) - \ln(V)$ being a 'fingerprint' of SCLC. The differential slope of I–V characteristics is defined as:

$$\alpha = \frac{d(\ln J)}{d(\ln V)} \tag{4.35}$$

In polycrystalline and amorphous dielectrics electron traps are not monoenergetic but they are distributed in the band gap of a material. Let us assume that there is the exponential energy distribution of traps:

$$N_t(E) = \frac{N_t}{E_o} \exp\left(-\frac{E - E_t}{E_o}\right) \tag{4.36}$$

Here N_t is the bulk concentration of traps in a dielectric, E_t is the energy level from which the distribution is exponential (relative to the bottom of the conduction band). E_o is the characteristic energy of distribution, given by the following equation:

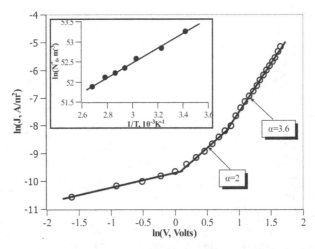

Figure 4.40. Typical I–V characteristic of the sample LN135-T in the double logarithmic coordinates at the temperature of 300 K. The inset demonstrates temperature dependence of trap concentration in the band gap of LiNbO$_3$.

$$E_o = lkT \tag{4.37}$$

In this case I–V characteristics are described in the following way [57]:

$$J = q^{(1-l)}\mu N_c \left(\frac{2l+1}{l+1}\right)^{(l+1)} \left(\frac{l\varepsilon\varepsilon_0}{(1+l)N_t^*}\right)^l \frac{V^{l+1}}{d^{2l+1}}. \tag{4.38}$$

Here q is the elementary charge, μ the carrier's mobility, N_c the effective density of states at the lower edge of the conduction band, $\varepsilon_0 = 8.85 \times 10^{-12}$ F m^{-1} is the electric constant, ε and d are the dielectric constant and thickness of a dielectric layer, respectively. Concentration N_t^* is given by:

$$N_t^* = N_t \exp\left(\frac{E_t}{lkT}\right) \tag{4.39}$$

The parameter l in equations (4.37) and (4.38) can be found as $l = \alpha_m - 1$, where α_m is the maximum magnitude of this parameter, given by equation (4.34).

The critical voltage of transition from an Ohmic to a 'quadratic' section of I–V curve is given by the following expression [57]:

$$V_x = \frac{qd^2 N_t^*}{\varepsilon\varepsilon_0} \left(\frac{n_0}{N_c}\right)^{1/l} \frac{l+1}{l} \left(\frac{l+1}{2l+1}\right)^{(l+1)/l} \tag{4.40}$$

Here n_0 is the concentration of free carriers. The critical voltage, corresponding to transition from the 'quadratic' law to the 'trap-filled-limit' law is defined as [57]:

$$V_{TFL} = \frac{qd^2}{\varepsilon\varepsilon_0} \left(\frac{9(N_t^*)^l((l+1)/l)^l((l+1)/(2l+1))^{l+1}}{8N_c}\right)^{1/(l-1)} \tag{4.41}$$

Using the magnitudes of V_x and V_{TFL}, determined from experimental I–V curves (figure 4.40) and the power exponent l, calculated through experimental parameter α, we solve equations (4.40) and (4.41) for N_t^* and n_0. The temperature dependence of $N_t^*(T)$ is a straight line in the $\ln(N_t^*) - 1/T$ coordinates (the inset in figure 4.40), which indicates the correctness of our assumption about the exponential distribution of traps. The magnitude of E_t is determined by the slope of a graph $\ln(N_t^*) - 1/T$, and N_t is obtained by extrapolation of this curve to $1/T \to 0$. We have determined the following magnitudes: $n_0 = 7 \times 10^{13}$ cm^{-3}, $N_t = 3.0 \times 10^{14}$ cm^{-3}, $E_t = 0.4$ eV. Distribution of traps in the band gap of the film is shown in figure 4.41.

Thus, the TA of Si–LiNbO$_3$ heterostructures, fabricated in an Ar(60%) + O$_2$(40%) gas mixture results in a decrease of bulk concentration of traps, which is in good agreement with other works [55, 56]. This decrease can explain the two times decline in coercive field of the sample LN135-T compared to as-grown heterostructures (see table 4.5). The results of [44] demonstrated a similar effect in the TA process of LiNbO$_3$ films in an oxygen atmosphere, and this phenomenon is explained by a decrease in concentration of oxygen vacancies, segregated at the grain boundaries.

The influence of the 'parasitic' LiNb$_3$O$_8$ phase, formed in the process of TA, on electrical phenomena in the studied heterostructures can be revealed by IS technique.

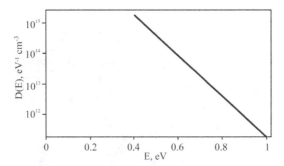

Figure 4.41. Distribution of traps in the band gap of LiNbO$_3$ film after TA.

4.3 Impedance spectroscopy of Si–LiNbO$_3$–Al heterostructures after thermal annealing

To make use of IS methods we choose the heterostructures with films having both LiNbO$_3$ and LiNb$_3$O$_8$ phases before TA (sample LN133). Nyquist diagrams and the frequency dependences of the imaginary parts of complex impedance and dielectric modulus for sample LN133-T are shown in figure 4.42. Analysis of IS spectra is conducted with the use of the equivalent circuit similar to sample LN133, shown in the inset of figure 4.17.

Nyquist diagrams, shown in figure 4.42(a), are almost ideal semicircles which indicate the single relaxation times rather than their distribution observed for the sample LN133. This means that TA leads to improved homogeneity of synthesized films in the studied heterostructures. By analysing $Z''(\omega)$ and $M''(\omega)$ spectra in the same way as for the sample LN133 and taking into account that TA results in the increase of average grain size up to $d_g = 80$ nm (see chapter 2 and [2]), we have obtained equal average magnitudes for both thicknesses of the bulk zone of a grain and its grain boundary $d_b \approx d_{gb} = 40$ nm ($d_b/d_{gb} = 1$) [22]. This fact is in good agreement with [2] where we have shown the increase in the amount of the LiNb$_3$O$_8$ phase in deposited films after TA. According to the model, proposed in [58], in the process of TA atmospheric oxygen penetrates into a film and neutralizes oxygen vacancies, reducing the migration ability of atoms. Evaporation of Li$_2$O occurs during the incubation period, producing the formation of niobium vacancies Nb$_v$. This process forces atoms to move in their neighborhood which results in nucleation of LiNb$_3$O$_8$–LiNbO$_3$ pair and phase separation. Thus, desorption of the Li$_2$O oxide is the limiting factor for crystallization rate. When nuclei LiNb$_3$O$_8$ and LiNbO$_3$ are formed, crystallization and phase transitions continue with a rate which depends on annealing parameters. Oxidation of Li, forming the light compound Li$_2$O, increases the loss of Li in a film which plays an important role in TA.

Figure 4.43 demonstrates the temperature dependence of conductivity (in the Arrhenius coordinates), affected by the grain boundaries $\sigma_{gb}(T)$ and bulk of grains $\sigma_b(T)$.

It follows from figure 4.43 that both components of conductivity are activated processes with activation energies of $E_{a1} = 0.3$ eV and $E_{a2} = 0.05$ eV for $\sigma_{gb}(T)$ and

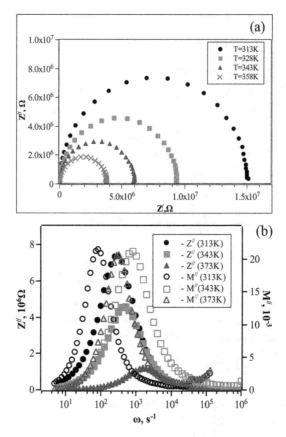

Figure 4.42. Nyquist diagrams (a) and frequency dependences of imaginary parts of complex impedance and dielectric modulus (b) for the sample LN133-T at different temperatures.

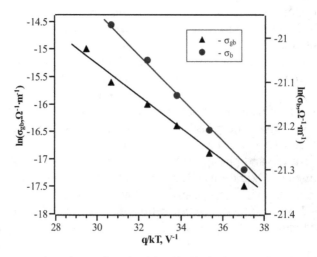

Figure 4.43. Temperature dependence of conductivity (the Arrhenius coordinates), associated with grain boundaries $\sigma_{gb}(T)$ and bulk of grains $\sigma_b(T)$ for the sample LN133-T.

$\sigma_b(T)$, respectively. Apparently, in the process of TA nonstoichiometric oxygen is accumulated at the grain boundaries being its source for the $LiNbO_3 \rightarrow LiNb_3O_8$ phase transition, described above, which decreases the height of an intergranular barrier slightly. As a result, the inter-granular region of grains (the $LiNb_3O_8$ phase) rises so much that grain boundaries and bulk regions play an equal role in conductivity, making the properties of films more 'uniform'. Figure 4.44 shows the frequency dependence of conductivity for the sample LN133-T at different temperatures.

Four segments corresponding to different power exponents in equation (4.20) and different conductivity mechanisms are observed on $\sigma(\omega)$ dependences (see figure LN133-T). The first segment at low frequency is affected by dc conductivity and as has been shown earlier its activation energy is 0.3 eV. This magnitude agrees with the activation energy, determined above for $\sigma_{gb}(T)$, which is the inter-granular barrier height. In region 2 in figure 4.4 conductivity is a thermally activated process with activation energy of 0.2 eV and determined from the slope of the Arrhenius graph, shown in figure 4.45 [22]. This magnitude is in agreement with those obtained for the sample LN133 earlier in the framework of the DCR model.

Figure 4.46 demonstrates the temperature dependence of the power exponent s in equation (4.20), attributed to the third and fourth regions in figure 4.44.

Temperature dependence of $s_2(T)$ is a typical one for conductivity, caused by the resonant absorption of quantums with energy $\hbar\omega$, corresponding to the energy of applied electric field $E(\omega)$ [59]. Since this mechanism is not observed for as-grown heterostructures (sample LN133), we can assume that it is caused by the increase in the amount of $LiNb_3O_8$ phase in films after TA. Region 4 in the frequency dependence $\sigma(\omega)$ (see figure 4.44) can be described in the framework of ADWP model. However, the parameter T_0 (see equation (4.21)) in this case is $T_0 = 1436$ K. Because this parameter describes dispersion of the bulk potential, apparently TA leads to re-distribution of defects in the bulk of grains.

Thus, TA results in a slight decrease of the intergranular barrier, increasing the role of bulk electrical conductivity which declines due to extension of the $LiNb_3O_8$ into the bulk areas of grains.

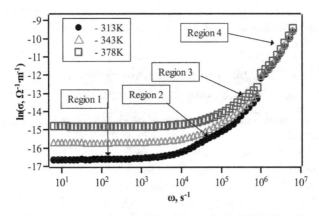

Figure 4.44. Frequency dependence of conductivity for sample the LN133-T at three temperatures.

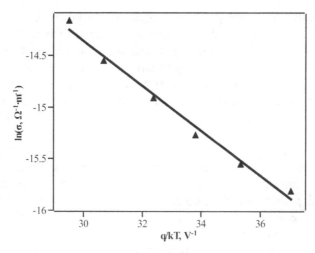

Figure 4.45. Temperature dependence of conductivity corresponding to region 2 in figure 4.44 in Arrhenius coordinates.

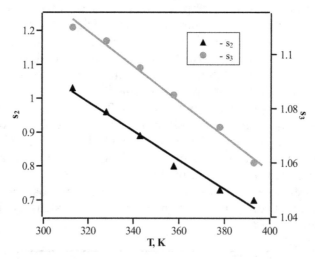

Figure 4.46. Temperature dependence of the power exponent s in equation (4.20) attributed to regions 3 and 4 in figure 4.44.

4.4 Optical band gap shift in thin LiNbO₃ films depending on RFMS conditions and subsequent thermal annealing

The band gap of materials in a semiconductor heterostructure, is an important parameter affecting its electrical and optical properties. In spite of the fact that optical properties of lithium niobate are well studied, recent publications have revealed variations in these properties. Specifically, the theoretically and experimentally obtained magnitude of the optical band gap varies from 3.57 eV [60] to 4.7 eV [61]. In fact, the band gaps of thin oxide films are influenced by many factors such as the size of polycrystalline grains [62], mechanical stress, caused by the mismatch of crystal lattice parameters of a substrate and a film [63], defect concentration (vacancies) [64] etc. Furthermore, as was shown in chapter 2 and [2, 56], TA being one of the most effective methods of post-deposition treatments, affects the structure and composition of LiNbO₃ films greatly. Taking into account a wide range of practical applications of LiNbO₃-based heterostructures in optoelectronics, the study of the effect of sputtering conditions on their optical properties is extremely important. Thin LiNbO₃ films were deposited by RFMS and IBS methods under the conditions given in table 4.7 (the shaded cells that represent the influencing parameter are used for comparison). Cleaved fluorphlogopite wafers were used as the transparent substrates for optical measurements. Subsequent TA was performed at the temperature of 650 °C for 60 min.

Single phase polycrystalline LiNbO₃ films fabricated under sputtering regimes given in table 4.7, have their grain size, texture and surface morphology influenced by the plasma properties [1, 2]. Thus, it is expected, that optical properties are strongly dependent on the synthesis regimes.

Figure 4.47 shows dependence of the adsorption coefficient on incident photon energy for studied LiNbO₃ films [65].

As can be seen from figure 4.47, initially the absorption coefficient α rises steadily with incident photon energy, and the sharp increase is observed in the energy range from 4 eV to 4.5 eV. The fundamental interband transition in single crystal lithium niobate is attributed to the valence-band maximum at the Γ point and the conduction-band minima at the 0.4Γ–K point of the Brillouin zone [66]. It is known, that both direct and indirect transitions are observed in single crystal LiNbO₃. The following equation describes the absorption coefficient α depending on the band structure of a semiconductor or a dielectric [67]:

$$\alpha(\nu) \propto B\left(h\nu - E_{\mathrm{g}}\right)^{r} \tag{4.42}$$

Here B is a frequency independent factor, h is Planck's constant, ν is the frequency, E_{g} the band gap, r a parameter ($r = 1/2$ for allowed direct transitions and $r = 2$ for allowed indirect ones). The segments of sharp increasing between 4.0 eV and 4.5 eV in figure 4.47, can be attributed to the allowed direct transitions.

As follows from equation (4.42) these segments should be the straight lines in the $\alpha^2 - h\nu$ coordinates. Figure 4.48 demonstrates a linear dependence of α^2 versus $h\nu$ graphs in the range of 4.0–4.5 eV, which indicates that the allowed direct transitions occur in this energy span [65].

The energy of the allowed direct transitions $E_{\mathrm{g}}^{\mathrm{dir}}$, obtained as an intercept of linear segments of $\alpha^2(h\nu)$ graphs with the horizontal axis are given in table 4.8.

The second segment, marked by 'B' in figure 4.47 for samples LN-2, LN-3 and LN-4 can be attributed to the allowed indirect transitions in the studied films, and according to equation (4.42) absorption spectra should be linear in $\alpha^{1/2} - h\nu$ coordinates as shown in figure 4.49.

Experimental data indicates that the direct band gap energy is of 3.60 and 3.68 eV for congruent and stoichiometric lithium niobate, respectively [68], which is close to the theoretically predicted magnitude (3.57 eV) [60]. However, taking into account

Table 4.7.

Table 4.7. Deposition regimes of samples for optical investigation.

Sample #	LN-1	LN-2	LN-3	LN-4	LN-4-T
Deposition technique	IBS	RFMS	RFMS	RFMS	RFMS
Magnetron power/supply power (W)	2000	100	100	100	100
Substrate–target distance (cm)	6	6	6	6	6
Substrate temperature (°C)	Unheated	Unheated	550	550	550
Substrate position	Offset from the target erosion zone	Offset from the target erosion zone	Over the target erosion zone	Offset from the target erosion zone	Offset from the target erosion zone
Annealing	–	–	–	–	+

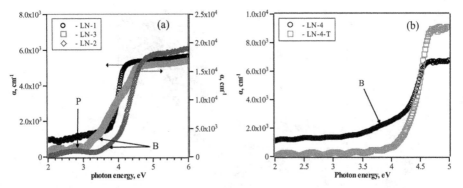

Figure 4.47. Dependence of the absorption coefficient α on the incident photon energy for samples LN-1, LN-2, LN-3 (a) and LN-4, LN-4-T (b) [65]. Reprinted with permission from Elsevier. Copyright 2017.

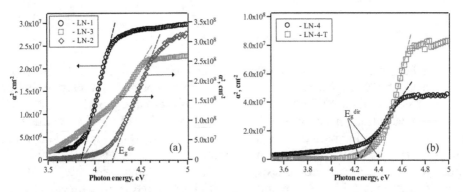

Figure 4.48. Dependence of the absorption coefficient α^2 on the incident photon energy for samples LN-1, LN-2, LN-3 (a) and LN-4, LN-4-T (b) [65]. Reprinted with permission from Elsevier. Copyright 2017.

Table 4.8. Direct, indirect band gap energies and concentration of CLC for studied samples.

Sample #	Direct energy gap, E_g^{dir} (eV)	Indirect energy gap, E_g^{indir} (eV)	Concentration of traps, N_t (cm^{-3})
LN-1	3.8	–	8×10^{19} [55]
LN-2	4.2	2.8	–
LN-3	3.8	2.5	7×10^{18} [25]
LN-4	4.2	2.0	2×10^{18} [55]
LN-4-T	4.4	–	3×10^{14} [55]

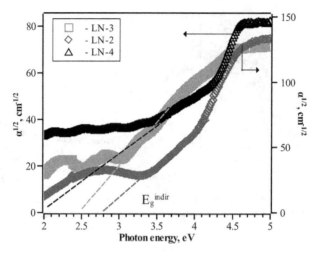

Figure 4.49. Dependence of $\alpha^{1/2}$ on incident photon energy for samples LN-2, LN-3, LN-4 [65]. Reprinted with permission from Elsevier. Copyright 2017.

electron–hole interaction and corrections applied to the existed models, the authors of [66] have obtained the magnitude of 4.7 eV. It was reported that interband optical transitions are limited by the presence of CLC in the band gap of a material [69]. Thus, the variation in energies E_g^{dir}, reflected in table 4.8, indicates the strong influence of sputtering parameters on the band gap of LiNbO$_3$ films and apparently can be attributed to the transition of carriers from these states to the bands. The structure and electrical properties of thin LiNbO$_3$ films, fabricated under the conditions given in table 4.7, are influenced by RFMS and IBS parameters [1, 2, 5]. It was revealed in chapter 2 and our work [1], that films, fabricated by the IBS method manifest a lower O/Nb ratio than films deposited by RFMS. This fact correlates with the results, given in table 4.8, when the sample LN-1, fabricated by IBS technique, demonstrates a narrower band gap (but higher concentration of CLC) than the sample LN-2, deposited by RFMS. As regards substrate temperature, it does not influence the direct bang gap E_g^{dir} in the studied films (see samples LN-2 and LN-4 in table 4.8).

In contrast, it follows from table 4.8 (samples LN-3 and LN-4), that the relative position of the substrate–target is a key parameter of RFMS, and has a profound effect on direct optical transitions in LiNbO$_3$ thin films. It was demonstrated in chapter 2 and our work [1], that all samples, located over the target erosion zone, are influenced by plasma (the ion assist effect), causing the formation of c-oriented LiNbO$_3$ films, compared to the films fabricated when they are offset from the target erosion zone and containing arbitrary oriented polycrystalline grains. Furthermore, the films, deposited under the same conditions as sample LN-3 (see table 4.7), have a relatively high concentration of traps (defects) ($N_t = 7 \times 10^{18}$ cm^{-3}) [25] compared to the films fabricated without the ion assist effect (sample LN-4) and having a trap concentration of $N_t = 2 \times 10^{18}$ cm^{-3} [55]. The intensive bombardment under the ion assist effect, causes the formation of a surface layer with high defect concentration, taking part in optical absorption and affecting the direct bang gap E_g^{dir}.

The effect of TA on the band gap, reflected in table 4.8, can have two origins: a decrease in the defect concentration and reduction of mechanical strain on the process of TA. Specifically, it was reported in [70], that the band gap of LiNbO₃ single crystals increases with decrease of defect concentration. Moreover, TA leads to a decline in mechanical strain, causing the blue shift of optical band gap in the annealed LiNbO₃ films [63]. Our results suggest (see table 4.2) that the direct band gap E_g^{dir} of LiNbO₃ films increases from 4.2 to 4.4 eV after TA (samples LN-4 and LN-4-T). This is in good agreement with our work [55] according to which TA results in decrease of concentration of CLC from 2×10^{18} cm^{-3} to 3×10^{14} cm^{-3} in the films grown and annealed under regimes similar to samples LN-4 and LN-4-T. The fact that TA leads to the disappearance of a texture in the films with the increase of the average grain size (see chapter 2 and [2, 56]) indicates the decline in mechanical strain in deposited LINbO₃ films. Furthermore, electrical conductivity and dielectric properties of annealed LiNbO₃ films are influenced by the bulk properties of grains rather than grain boundaries [22], making their optical properties close to one of single crystal lithium niobate.

It was shown in [71], that the absorption edge below 3.8 eV has an indirect nature in single crystal LiNbO₃. It is generally accepted, that indirect transition is a transition with suitable photon participation when the bottom of the conduction zone and the top of the valence band are not at the same wave vector K as observed in bulk lithium niobate. As shown, LiNbO₃ films demonstrate high defect concentration creating band tails in the band gap, with the possibility of indirect band–band tail transitions. The results of [71] suggest, that the band tails are caused by electric fields, produced by oxygen vacancies and produce indirect optical transitions with energy of 3.5 eV. We have demonstrated that TA of as-grown LiNbO₃ films results in a decrease of defect concentration (oxygen vacancies or antisite defects Nb_{Li}^{+4}) from 1×10^{18} cm^{-3} to 5×10^{16} cm^{-3} [2]. However, the indirect band gap energies in our experiments were lower than 3.5 eV (see table 4.8). There is some evidence, that such low energetic transitions can be associated with polaron hopping conductivity, affecting the absorption coefficient [72]. The absorption band at energy E_{opt} can be expressed as:

$$E_{opt} = 4W \qquad (4.43)$$

where W is the activation energy of conductivity.

It is important to note, that the peak with energy of $E_p \approx 2.8$ eV, marked as 'P' in figure 4.47(a), has been reported by other investigators. They stress that the amplitude of a broad peak, observed at energy of 2.5 eV in their absorption spectra $\alpha(\omega)$, depends on the degree of reduction in LiNbO₃ single crystals [64] and attributed to polaron hopping conductivity. Based on the detailed analysis of various polaronic models, another group of authors concluded that a broad peak in the absorption coefficient $\alpha(\omega)$ changes its position in the range of low energies depending on the particular type of polarons taking part in optical absorption via hopping conductivity. There are the following relationships between the observed peaks and different types of polarons: a peak with energy of 0.94 eV—free small polarons, 1.64 eV—bound polarons, 2.50 eV—bound bipolarons [38]. Since the energy of the peak 'P' in spectra $\alpha(\omega)$ from samples LN-2 and LN-3 (see figure 4.18) is close to 2.5 eV, we can conclude, that in our samples the bound bipolarons are responsible for optical absorption at this energy magnitude.

4.5 Temperature transition of p- to n-type conduction in the LiNbO$_3$/ Nb$_2$O$_5$ polycrystalline films fabricated in an Ar + O$_2$ reactive gas environment

One of the interesting issues that we would like to discuss at the end of this chapter is the effect of temperature transition of p- to n-type conduction in LiNbO$_3$ films. The study of this phenomenon is a separate and in-depth piece of research, but we have attempted to explain it. Thin films (300 nm) were fabricated by the RFMS method without the ion assist effect. Sputtering was performed at a magnetron power of 100 W in an Ar atmosphere and in an Ar+O$_2$ gas mixture with the ratio of Ar/O$_2$ = 60/40 and 80/20 ($P = 1.5 \times 10^{-1}$ Pa). The substrates were located offset from the target erosion zone at a distance of 5 cm from the target. Silicon wafers (001)Si of n-type conductivity ($\rho = 7.2$ Ohm cm) were used as substrates. For the film synthesis process, the substrates were heated up to a temperature of 550 °C.

Figure 4.50 shows XRD patterns of the studied films with the thickness of 300 nm.

All XRD patterns contain peaks attributed to the rhombohedral cell of LiNO$_3$. Peaks, associated with the angles of $2\Theta = 24.3°$ ($\bar{3}11$) and $2\Theta = 30.3°$ ($\bar{1}11$) at the spectra 2 and 3 (see figure 4.50) correspond to the monoclinic lattice of Nb$_2$O$_5$ ($a = 12.74$ Å, $b = 4.88$ Å, $c = 5.56$ Å). The relative intensity of the peaks, attributed to the LiNbO3 phase in figure 4.50 corresponds to the intensity of peaks for

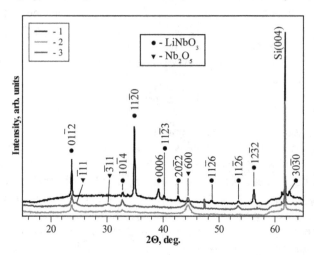

Figure 4.50. XRD patterns of studied films, fabricated by RFMS method in different reactive gas environments: 1 pure Ar, 2 Ar + O$_2$ gas mixture with Ar/O$_2$ = 60/40, 3 Ar + O$_2$ gas mixture with Ar/O$_2$ = 80/20.

polycrystalline lithium niobate from the crystallographic database (see appendix), which indicates the arbitrary grain orientation in all studied films. Some of the Nb_2O_5 grains are oriented randomly (peaks $(\bar{3}11)$ and $(\bar{1}11)$), whereas others correlate to a single-axis texture (the crystallographic plane (600) is parallel to the substrate plane). The quantitative ratio of the phases $LiNbO_3$ and Nb_2O_5 was estimated using the reference intensity ratio (RIR) for non-textured phases. The mass fractions of the $LiNbO_3$ phase and non-textured Nb_2O_5 in the film, deposited in an Ar + O_2 gas environment with ratio Ar/O_2 = 80/20, is about 65% and 35%, respectively. In films fabricated in an Ar + O_2 gas mixture with ratio Ar/O_2 = 60/40, it is around 55% and 45%, respectively. Taking into account these fractions we can conclude that the amount of non-textured Nb_2O_5 phase in the film declines by 20%, whereas the amount of the textured Nb_2O_5 phase (based on the intensity of (600) peaks) decreases four-fold when the oxygen content reduces from 40% to 20% in a reactive plasma.

The grain size, determined from the XRD patterns (the size of the coherent scattering areas) is $l = 21$ nm and $l = 10$ nm for $LiNbO_3$ and Nb_2O_5 grains, respectively [73]. This result coincides with the TEM and micro diffraction patterns shown in figures 4.51 and 4.52.

Figure 4.51 presents TEM micro diffraction pattern (a) and a bright-field TEM image (b) of the film with a thickness of 100 nm, deposited by RFMS in an Ar atmosphere. All reflexes in the micro diffraction pattern are attributed to $LiNbO_3$ phase with arbitrary grain orientation. The bright-field TEM pattern (figure 4.51(b)) indicates that the film contains grains sized from 50 to 100 nm.

Figure 4.52 shows TEM micro diffraction pattern (a), bright-field (b) and dark-field (c, d) images of the thin (100 nm) film deposited by RFMS method in an Ar + O_2 environment at the Ar/O_2 ratio of 60/40. All reflexes in this pattern correspond to the $LiNbO_3$ and Nb_2O_5 phases with randomly oriented grains. It follows from the analysis of TEM patterns, that the films, fabricated in this regime contain single-phase crystalline $LiNbO_3$ blocks with the size of 50–100 nm and the same size blocks

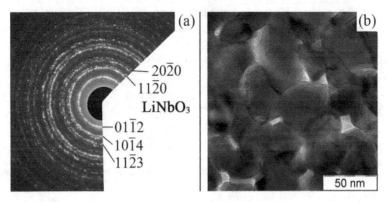

Figure 4.51. TEM diffraction pattern (a) and a bright-field (b) TEM image of the thin (100 nm) film deposited by RFMS method in an Ar atmosphere [73]. Reprinted from with permission from Elsevier. Copyright 2017.

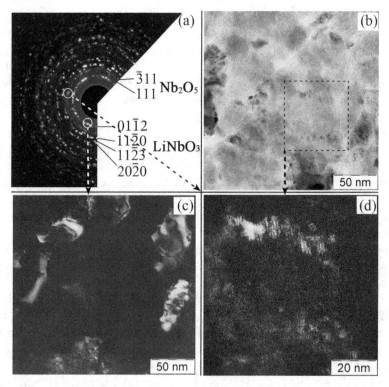

Figure 4.52. TEM micro diffraction pattern (a), bright-field (b) and dark-field (c, d) images of the thin (100 nm) film deposited by RFMS method in an Ar + O_2 environment at the Ar/O_2 ratio of 60/40 [73]. Reprinted with permission from Elsevier. Copyright 2017.

composed of LiNbO$_3$ sub-grains and nano-grains of Nb$_2$O$_5$ (figure 4.52(b)). The circled area in figure 4.52(b) illustrates the nano-grains of Nb$_2$O$_5$ sized from 5 to 10 nm within a single LiNbO$_3$ block. Their size was estimated based on the total sum of intensities $(\bar{3}11)$ and $(\bar{1}11)$ (figure 4.52(d)). Furthermore the detailed analysis of TEM images reveals that some blocks are composed only of the nano-grains of Nb$_2$O$_5$ (see figure 4.53).

According to the mechanism proposed in [74], the Nb$_2$O$_5$ oxide is formed as a product of the following dissociation: $2\text{LiNb}_3\text{O}_8 \rightarrow 3\text{Nb}_2\text{O}_5 + \text{Li}_2\text{O}$.

Since LiNbO$_3$-based heterostructures, fabricated in an Ar + O_2 reactive gas environment can be represented as two back-to-back Schottky barriers connected in series, it makes the charge transport from the film into a Si substrate at low applied voltage energetically unfavorable. Thus, during the Hall measurements a conducting substrate has no contribution.

It is known, that if only one type of carrier is presented in a material the Hall coefficient (RH) is inversely proportional to the free-carrier concentration (n for electrons or p for holes) according to the following expressions [75]:

$$R_{\mathrm{H}} = \frac{1}{q \cdot p} r \text{ - for a p-type semiconductor}$$

$$R_{\mathrm{H}} = \frac{1}{q \cdot n} r \text{ - for a n-type semiconductor}$$

(4.44)

Here p and n are the concentrations of holes and electrons, respectively, r is the scattering factor defined as $r = \mu_{\mathrm{H}}/\mu_{\mathrm{d}}$ (μ_{H} and μ_{d} the Hall and drift mobility, respectively). The sign of R_{H} corresponds to the sign of the major charges in a semiconductor: $R_{\mathrm{H}} > 0$ for a p-type semiconductor, whereas $R_{\mathrm{H}} < 0$ for an n-type semiconductor. Equations (4.44) are used to determine free-carrier concentration based on the experimental Hall coefficient R_{H}. Traditionally, the scattering coefficient lies in the range from 1 for metals to 2 for semiconductors. The Hall coefficient in the films, fabricated by RFMS in an Ar + O_2 environment with the ratio Ar/O_2 = 60/40 has a complex temperature dependence, which is demonstrated in the Arrhenius coordinates in figure 4.54.

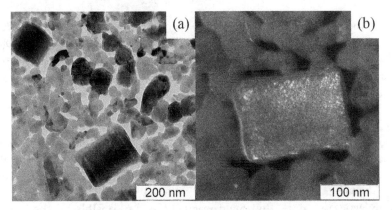

Figure 4.53. TEM patterns of the LiNbO$_3$/Nb$_2$O$_5$ thin (100 nm) film fabricated by RFMS method in an Ar + O_2 environment at the Ar/O_2 ratio of 60/40. Patterns (a) and (b) demonstrate the single crystalline blocks composed entirely of Nb$_2$O$_5$ nano-grains [73]. Reprinted with permission from Elsevier. Copyright 2017.

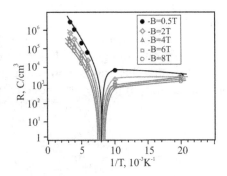

Figure 4.54. Temperature dependence of the Hall coefficient at different applied magnetic fields B for the films, fabricated by RFMS in an Ar + O_2 atmosphere at the Ar/O_2 = 60/40 ratio.

As seen from figure 4.54, R_H is almost temperature independent at low temperature and decreases sharply when $T > 100$ K, becoming negative at temperature of 150 K. After that, the Hall coefficient rises sharply, changing its sign from negative to positive and monotonically increases at temperatures close to 300 K.

Interpretation of the Hall measurements in polycrystalline semiconductors becomes difficult due to the presence of the trapped charge at the interfaces. Grain boundaries contain high concentration of CLC (border traps) capturing carriers from a bulk or scattering them. The maximum charge which can be captured depends on both surface density (N_t) and energy position (E_t) of traps, influencing activation energy. The depletion zone can be smaller than the average grain size or extended through the entire volume, so the carrier free path can be lower or higher than the size of a grain. Based on the model, proposed in [75], when the depletion zone extends only partially into a grain, the free-carrier concentration, derived from the Hall measurements, gives their bulk concentration. The Hall mobility is given by [75]:

$$\mu_b = \mu_0 \exp\left(-\frac{\varphi_b}{kT}\right) \tag{4.45}$$

where φ_b is the band bending at the grain boundaries, μ_0 is a parameter, dependent on the grain size l as $\mu_0 \approx 10^6 l$ at room temperatures (l is measured in m, μ_0—in m^2 V^{-1} s^{-1}). The band bending, associated with the doping level N, reaches a maximum when the trap density $N_t = Nl$, and the mobility is minimal.

Since the Hall coefficient is almost temperature independent in the temperature range of $T = 50$–100 K, the barrier height φ_b is low and the depletion area width is small compared to the grain size. Thus, the carrier mobility in this temperature range is limited by the bulk properties of grains, and concentration of free carriers, determined from the Hall measurements, is the concentration in a bulk of grains. Concentration of major carriers (holes) in this temperature range, estimated using equation (4.44) has a magnitude of $N_p = 2.2 \times 10^{15}$ cm^{-3}.

Some investigators note, that oxygen vacancies, responsible for the formation of hole polarons, can be also responsible for p-type conductivity in LiNbO$_3$ [76]. In contrast, LiNbO$_3$ films, deposited under the same conditions, but at lower partial oxygen pressure, manifest n-type conductivity with high donor concentration [77]. The change in sign of the Hall coefficient observed in our experiments can occur when two types of carriers (electrons and holes) exist in a film. At low temperature, when the concentration of holes exceeds those for electrons ($p \gg n$) p-type conductivity takes place and the Hall coefficient is positive. The observed decrease at the temperature dependence shown in figure 4.54 at $T > 100$ K can be produced by emission of electrons from interface states, contributing total conductivity at these temperatures. The change of p- to n-type of conductivity occurs due to higher mobility of electrons compared to holes. In this situation the Hall coefficient is given by the following equation [78]:

$$R_H = \frac{1 - xb^2}{qp(1 + xb)^2} r \tag{4.46}$$

where b is a parameter which is equal to the ratio of electron and hole mobilities μ_n/μ_p, $x = n/p$, $r = \mu_H/\mu_d$ is a scattering factor. The Hall coefficient becomes zero when electron and hole contributions compensate each other $(1 = xb^2)$ and goes through the minimum when $x = (b + 2)/b^2$. The minimal magnitude of R_H is:

$$R_{min} = -\frac{1}{4}\frac{b^2 \cdot r}{(b + 1)pq} \tag{4.47}$$

Thus:

$$\left| \frac{R_{max}}{R_{min}} \right| = \frac{4(b + 1)}{b^2} \tag{4.48}$$

Here R_{max} is the maxima Hall coefficient at low temperatures, given by equation (4.44), when only holes take part in conductivity. Solving equation (4.48) for b and using experimental magnitudes of R_{max} and R_{min}, we obtain the parameter $b = 49$ for the studied films.

Furthermore, the Nb_2O_5 phase can contribute n-type conductivity at these temperatures. It was demonstrated in [79], that n-type conductivity of Nb_2O_5 is extremely sensitive to the partial pressure of oxygen in the synthesis process. Oxygen vacancies $V_O^{2\bullet}$ and cation interstitials are able to capture electrons, generating deep traps in the band gap of Nb_2O_5, causing its n-type conductivity. Conductivity of this oxide is affected by the partial oxygen pressure P_{O_2} via the dependence of concentration of oxygen vacancies according to the law: $\sigma = const \cdot P_{O_2}^{-0.25}$ [79]. Thus, the change in the Ar/O_2 ratio in the reactive chamber is an effective tool for fabrication of $LiNbO_3/Nb_2O_5$ films with the desired electrical properties.

In polycrystalline materials grain boundaries play an additional and often a leading role in the charge transport phenomena. As a result, the increase in the Hall coefficient with temperature in the range of 150–300 K can be attributed to the inter-granular barriers at the interfaces. It was demonstrated in [80] that mobility rises with temperature in a bulk material and declines with temperature in thin films. Mobility is a rising function of temperature according to equation (4.45), provided the charge transport in polycrystalline films is limited by inter-granular barriers φ_b. In the temperature range of 200–300 K the Hall coefficient is influenced by inter-granular barriers and the scattering at interfaces. There are three main scattering mechanisms in thin films: surface scattering, dislocation scattering and grain boundary scattering. Since the thickness of the studied films (300 nm) exceeds the mean free path for carriers, the surface scattering does not influence the charge transport in our case. Evidently, the grain boundary scattering at the barriers with height of φ_b is a dominant scattering mechanism. In [75] the following equation for a maximum potential barrier height was proposed when $Nl = N_t$:

$$\varphi_b^{\ max} = \frac{q^2 N_t l}{8\varepsilon\varepsilon_0} \tag{4.49}$$

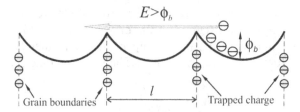

Figure 4.55. Schematic representation of a polycrystalline n-type semiconductor when the mean free path exceeds the average grain size l [73].

Here N_t is the trap density and l the average grain size. Based on the model, developed in [75], only electrons that have energy higher than the potential barrier height φ_b, are able to overcome few grains without a considerable scattering as schematically shown in figure 4.55.

Only electrons with energy of $E > \varphi_b$ are deflected by applied magnetic field, so the concentration, determined through the Hall measurements differs from the bulk one and can be expressed as:

$$n = n_1 \cdot \exp\left(-\frac{\varphi_b}{kT}\right) \qquad (4.50)$$

Thus, in the case when charge transport is limited by inter-granular barriers, temperature dependence of concentration derived from the Hall effect should be a linear function in the Arrhenius coordinates with a slope of φ_b. As can clearly be seen from figure 4.54, the Hall coefficient is a thermally activated process with activation energy of $\varphi_b = 0.13$ eV in the temperature range of 200–300 K. Since the average grain size in the studied films is $l = 20$ nm and taking into account that $\varepsilon = 28$ we estimated the density of border traps using equation (4.49) and have obtained the following magnitude: $N_t = 8.0 \times 10^{12}$ cm^{-2}. Another contribution in this activated process can be associated with thermal activation of electrons from traps in Nb_2O_5 with activation energy of 0.15 eV [79].

The scattering factor (or the Hall factor) $r = \mu_H/\mu_d$ is used to describe the scattering of carriers. Thus, the magnetic field dependence of the Hall coefficient is determined by the respective function for the Hall factor $r(B)$. It was proposed to determine the Hall factor from experimental measurements, using the following expression:

$$r = \frac{R(0)}{R(\infty)} \qquad (4.51)$$

Here $R(0)$ and $R(\infty)$ are the Hall coefficient, recorded at low and strong magnetic fields, respectively.

Figure 4.56 demonstrates the field dependence of the Hall factor $r(B)$ for the studied films in the double logarithmic scale $\ln(r) - \ln(B)$.

As can be seen from figure 4.56, the Hall factor approaches 1 at strong applied magnetic fields and does not depend on temperature in full agreement with theory [54, 80]. Furthermore, it follows from figure 4.56 that the Hall factor depends on the

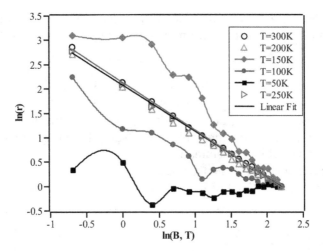

Figure 4.56. Dependence of the Hall factor of the studied $LiNbO_3/Nb_2O_5$ films on applied magnetic fields at different temperatures.

applied field in different ways depending on temperature. The Hall factor r can be of various magnitudes depending on the scattering mechanisms occurring in a material. Specifically, in bulk semiconductors it can be $r = 1.18$ for the thermal phonon scattering or $r = 1.93$ for the ionized impurities scattering [54]. In polycrystalline semiconductors when other scattering mechanisms dominate, the Hall factor, being a function of energy, can have magnitudes that differ from those for the bulk materials. The field dependence of the Hall factor in the temperature range of 200–300 K is given by:

$$r = A(T) \cdot B^s \qquad (4.52)$$

Here B is the magnetic field and parameters $A(T)$ and s can be found as an intercept of an experimental graph $\ln(r) - \ln(B)$ and its slope, respectively. The temperature dependence of the parameter $A(T)$ in equation (4.52) is a straight line in the Arrhenius coordinates as seen from figure 4.57 $E_a = 0.03$ eV, obtained from the slope of this graph [73]. This magnitude can be interpreted as a difference in activation energies for the Hall and drift mobilities.

As regards the power exponent s, it is temperature independent and is equal to $s = -1$.

As can be seen from figure 4.56, the field dependence of the Hall factor has a complex character in the range of 50–200 K. Earlier, based on the analysis of the Hall coefficient $R_H(T)$, it was demonstrated that carrier mobility is limited by the hole mobility in the bulk of grains at temperature of 50 K. In the range of low magnetic fields the Hall coefficient obeys the following law [54, 80]:

$$R_H(B) = R_H(0)[1 - \alpha \cdot \mu^2 B^2] \qquad (4.53)$$

Here $R_H(0)$ is the Hall coefficient at zero field, α is the non-linearity factor dependent on the scattering mechanism. Specifically, $\alpha = 0.99$ in the case of acoustic phonons,

$\alpha = 0.49$ for the ionized impurities scattering and $\alpha = 0$ for the neutral impurities scattering. On the other hand, at strong magnetic fields the Hall coefficient is given by [54, 80]:

$$R_H(B) = R_H(\infty)\left[1 + \frac{\alpha'}{\mu^2 B^2}\right] \qquad (4.54)$$

Here $R_H(\infty)$ is the Hall coefficient at strong magnetic fields (the high field limit), α' is the non-linearity factor in the high field limit. Note, that at low magnetic fields the carriers with higher mobility dominate.

Figure 4.58 demonstrates a field dependence of the Hall coefficient for the studied LiNbO₃/Nb₂O₅ films at $T = 150$ K [73].

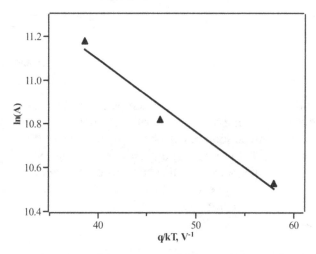

Figure 4.57. Temperature dependence of the pre-exponential factor $A(T)$ in equation (4.52) for the LiNbO₃/Nb₂O₅ films in the temperature range of 200–300 K in Arrhenius coordinates.

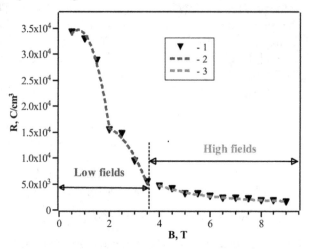

Figure 4.58. Field dependence of the Hall coefficient for the studied LiNbO₃/Nb₂O₅ films at temperature of 150 K. 1—experiment, 2—theoretically calculated using equation (4.53), 3—theoretically calculated using equation (4.54).

As was demonstrated in [54], if several types of carriers coexist in a semiconductor and their mobilities differ significantly, the field dependence $R_H(B)$ manifests as many points of inflection (quadratic segments) as many types of carriers contribute the Hall measurements. Two quadratic sections $R_{H1}(0)$ and $R_{H2}(0)$, observed at low fields in figure 4.58 are attributed to two types of carriers in full agreement with equation (4.53). Electrons with high mobility (contributing R_H at lower fields) and holes with less mobility (contributing R_H at higher fields) can play this role. Assuming that the same scattering mechanism occurs, we have obtained the following magnitudes using equation (4.53): $N_n = 1.8 \times 10^{14}$ cm^{-3}, $N_p = 3.9 \times 10^{14}$ cm^{-3}, $\mu_{Hn} = 0.6$ m^2 V^{-1} s^{-1}, $\mu_{Hp} = 0.45$ m^2 V^{-1} s^{-1} (concentration of carriers has been estimated from $R_{H1}(0)$ and $R_{H2}(0)$ through equation (4.46)) [73]. At high magnetic fields (see figure 4.58) the Hall coefficient decreases with field according to equation (4.54). Taking $\alpha' = 1$ we have calculated the Hall coefficient ($R_H(\infty) = 966$ C cm^{-3}), the Hall mobility ($\mu_{Hp} = 0.13$ m^2 V^{-1} s^{-1}), and concentration of holes ($N_p = 6.5 \times 10^{15}$ cm^{-3}) which are in a good agreement with work [81].

Summary and emphasis

1. Positive oxide charge in LiNbO$_3$ films, fabricated by RFMS in an Ar reactive gas environment, does not depend on working pressure or the presence of the LiNb$_3$O$_8$ phase and is not produced by defects created by the bombardment. The most likely source of this charge could be antisite defects Nb$_{Li}^{4\bullet}$ and oxygen vacancies.

2. The ferroelectric <0001>-textured LiNbO$_3$ films, deposited by RFMS with the ion assist effect, demonstrate the magnitude of remnant polarization close to the bulk lithium niobate regardless of reactive plasma composition. The increase in reactive gas pressure leads to a decrease in the coercive field, which is associated with the random grain orientation in the films, fabricated under these conditions.

3. The non-ferroelectric LiNb$_3$O$_8$ phase, existing in the films, fabricated at higher Ar pressure in a reactive chamber, is formed at the grain boundaries, whereas for the bulk grains it is composed of the LiNbO$_3$ phase. The ratio between the grain boundaries and bulk core widths is $d_{gb}/d_b = 0.4$. Each phase contributes electrical conductivity.

4. Polycrystalline LiNbO$_3$ films with an inter-granular barrier of 0.4 eV are formed in the process of RFMS regardless of the reactive gas pressure. The barrier height does not depend on the presence of the LiNb$_3$O$_8$ phase in synthesized films.

5. Thin LiNbO$_3$ films, deposited by RFMS in an Ar + O$_2$ (Ar/O$_2$ = 60/40) reactive gas mixture, have the lowest built-in field, which can be explained in terms of decrease in concentration of oxygen vacancies. The presence of oxygen atoms in a reactive chamber decreases concentration of lithium vacancies in the film and the probability of the formation of positively charged complexes. Lithium and oxygen atoms diffuse into the silicon substrate forming deep donors in Si–LiNbO$_3$ heterostructures.

6. Both DC and AC conductivity mechanisms along with dielectric relaxation mechanisms have been studied in detail for $LiNbO_3$-based heterostructures, fabricated at different RFMS conditions. Electrical properties of Si–$LiNbO_3$ heterostructures, synthesized in an Ar atmosphere, are influenced by CLC (traps) in the band gap of $LiNbO_3$ with activation energy of E_a = 0.7 eV (attributed to antisite defects $Nb_{Li}^{4\bullet}$) and E_a = 0.9 eV (attributed to the lithium vacancies). The presence of oxygen in a reactive chamber in the ratio of Ar/O_2 = 60/40 leads to the disappearance of the center with energy of 0.9 eV. At the same time, the extended intermediate defect layer at the Si/$LiNbO_3$ interface in heterostructures, fabricated in an Ar + O_2 environment, results in a high density of CLC, influencing electrical properties of the studied heterostructures.

7. As-grown $LiNbO_3$ films do not manifest preferable domain orientation. The films after TA demonstrate preferably oriented domains, which makes them non-symmetric in terms of polarization reversal relative to applied electric field. This effect does not depend on the sputtering regimes.

8. TA of as-grown $LiNbO_3$ films leads to decrease in the magnitude of positive effective charge, which is attributed to neutralization of oxygen vacancies in them. Furthermore, TA results in decrease of the interface states at the film/substrate interface, making Si–$LiNbO_3$ heterostructures close to MIS structures in the flat band regime.

9. TA has a positive effect on ferroelectric properties of $LiNbO_3$ films, decreasing built-in fields, which are responsible for the asymmetry of P–E loops. The decrease of trap concentration in the films after TA results in a decline of coercive field, especially for those fabricated in an Ar + O_2 gas mixture.

10. Hysteresis of C–V characteristics for Si–$LiNbO_3$ heterostructures after TA is explained in terms of the generation of holes and their subsequent re-emission to border traps on the Si/$LiNbO_3$ heterointerface. The generation of electron–hole pairs in the SiO_2 intermediate layer is caused by the Fowler–Nordheim emission of electrons during electrical measurements.

11. TA does not change the energy position of traps with E_t = 1.7 eV in the band gap of $LiNbO_3$. However, TA leads to re-crystallization of the films, resulting in an increase of polycrystalline grains and minimizing the inter-granular barrier effect on charge transport. Thus, conductivity of the studied films in a wide range of electrical fields is limited by thermal phonon scattering of electrons as for $LiNbO_3$ single crystals. It indicates the high efficiency of thermal annealing in terms of improvement of dielectric properties of $LiNbO_3$-based heterostructures.

12. Optical properties (optical absorption band edge) of thin $LiNbO_3$ films, fabricated by IBS and RFMS methods are very sensitive to the sputtering regimes. The studied films demonstrate both direct optical transitions with energies E_g^{dir} = (3.8–4.2) eV and indirect ones depending on the technological sputtering regimes. $LiNbO_3$ films, deposited under the ion assist

effect demonstrate the smallest direct band E_g^{dir} due to high concentration of CLC and mechanical strain, produced by the strong plasma influence on the film surface. Indirect optical transitions are caused by the band 'tails', generated by defects in fabricated films. TA of as-grown LiNbO$_3$ films leads to decrease in concentration of CLC and consequently to increase in the band gap up to $E_g = 4.4$ eV, which is close to those for single crystal LiNbO$_3$.

13. The films, deposited by RFMS without the ion assist effect in an Ar + O$_2$ atmosphere, contain two phases: LiNbO$_3$ and Nb$_2$O$_5$. The number of phases depends on the Ar/O$_2$ ration in a reactive chamber and the maximum phase ratio LiNbO$_3$(55%)/Nb$_2$O$_5$(45%) is observed for the reactive plasma composition of Ar/O$_2$ = 60/40. The films are composed of crystalline blocks, consisting of 20 nm sub-grains of LiNbO$_3$ and 10 nm nano-grains of Nb$_2$O$_5$. For the first time the temperature transition of p- to n-type conduction in LiNbO$_3$/Nb$_2$O$_5$ films has been observed. The ratio of drift mobilities for electrons and holes is $\mu_n/\mu_p = 49$. At high magnetic fields, holes with concentration of $N_p = 6.5 \times 10^{15}$ cm^{-3} play a crucial role in the Hall measurements. When temperature rises, electrons, trapped in LiNbO$_3$ and also attributed to n-type conductivity of the Nb$_2$O$_5$ phase, contribute conductivity, limited by the grain boundary scattering, associated with an inter-granular barrier of 0.13 eV. The effect of conductivity type change depends on the Ar/O$_2$ ratio in a reactive chamber and requires a more detailed study.

References

[1] Sumets M, Kostyuchenko A, Ievlev V, Kannykin S and Dybov V 2015 Sputtering condition effect on structure and properties of LiNbO$_3$ films *J. Mater. Sci. Mater. Electron.* **26** 4250–6

[2] Sumets M, Kostyuchenko A, Ievlev V, Kannykin S and Dybov V 2015 Influence of thermal annealing on structural properties and oxide charge of LiNbO$_3$ films *J. Mater. Sci. Mater. Electron.* **26** 7853–9

[3] Soergel E 2011 Piezoresponse force microscopy (PFM) *J. Phys. D. Appl. Phys.* **44** 464003

[4] Sumets M, Ievlev V, Kostyuchenko A, Vakhtel V, Kannykin S and Kobzev A 2014 Electrical properties of Si-LiNbO$_3$ heterostructures grown by radio-frequency magnetron sputtering in an Ar + O$_2$ environment *Thin Solid Films* **552** 32–8

[5] Sumets M, Ievlev V, Kostyuchenko A, Kuz'mina V and Kuzmina V 2014 Influence sputtering conditions on electrical characteristics of Si-LiNbO$_3$ heterostructures formed by radio-frequency magnetron sputtering *Mol. Cryst. Liq. Cryst.* **603** 202–15

[6] Hao L Z, Zhu J, Luo W B, Zeng H Z, Li Y R and Zhang Y 2010 Electron trap memory characteristics of LiNbO$_3$ film/AlGaN/GaN heterostructure *Appl. Phys. Lett.* **96** 32103

[7] Shandilya S, Tomar M and Gupta V 2012 Deposition of stress free c-axis oriented LiNbO$_3$ thin film grown on (002) ZnO coated Si substrate *J. Appl. Phys.* **111** 10–6

[8] Gordillo-Vázquez F J and Afonso C N 2002 Influence of Ar and O2 atmospheres on the Li atom concentration in the plasma produced by laser ablation of LiNbO$_3$ *J. Appl. Phys.* **92** 7651

[9] Smyth D M 1983 Defects and transport in LiNbO$_3$ *Ferroelectrics* **50** 93–102

[10] van Opdorp C and Kanerva H K J 1967 Current-voltage characteristics and capacitance of isotype heterojunctions *Solid State Electron.* **10** 401–21

[11] Sze S M and Kwok K N 2006 *Physics of Semiconductor Devices* (New York: Wiley)

[12] Gosele U and Tan T Y 1982 Oxygen diffusion and thermal donor formation in silicon *Appl. Phys. A Solids Surfaces* **28** 79–92

[13] Gosele U M 1988 Fast diffusion in semiconductors *Annu. Rev. Mater. Sci.* **18** 257–82

[14] Iyevlev V, Kostyuchenko A, Sumets M and Vakhtel V 2011 Electrical and structural properties of LiNbO$_3$ films, grown by RF magnetron sputtering *J. Mater. Sci. Mater. Electron.* **22** 1258–63

[15] Choi S-W, Choi Y-S, Lim D-G, Moon S-I, Kim S-H, Jang B-S and Yi J 2000 Effect of RTA treatment on LiNbO$_3$ MFS memory capacitors *Korean J. Ceram.* **6** 138–42

[16] Wemple S H, DiDomenico M and Camlibel I 1968 Relationship between linear and quadratic electro-optic coefficiens in LiNbO$_3$, LiTaO$_3$, and other oxygen-octahedra ferroelectrics based on direct measurement of spontaneous polarization *Appl. Phys. Lett.* **12** 209–11

[17] Kim S, Gopalan V and Gruverman A 2002 Coercive fields in ferroelectrics: A case study in lithium niobate and lithium tantalate *Appl. Phys. Lett.* **80** 2740–2

[18] Zubko P, Jung D J and Scott J F 2006 Space charge effects in ferroelectric thin films *J. Appl. Phys.* **100** 114112

[19] Maissel L I and Glang R 1970 *Handbook of Thin Film Technology* (New York: McGraw-Hill)

[20] Saxena A N 1969 Forward current-voltage characteristics of Schottky barriers on n-type silicon *Surf. Sci.* **13** 151–71

[21] Kashirina N I and Lakhno V D 2010 Large-radius bipolaron and the polaron–polaron interaction *Physics-Uspekhi* **53** 431–53

[22] Sumets M, Kostyuchenko A, Ievlev V and Dybov V 2016 Electrical properties of phase formation in LiNbO$_3$ films grown by radio-frequency magnetron sputtering method *J. Mater. Sci. Mater. Electron.* **27** 7979–86

[23] Akazawa H and Shimada M 2006 Precipitation kinetics of LiNbO$_3$ and LiNb$_3$O$_8$ crystalline phases in thermally annealed amorphous LiNbO$_3$ thin films *Phys. Status Solidi* **203** 2823–7

[24] Rost T A, Lin H, Rabson T A, Baumann R C and Callahan D L 1992 Deposition and analysis of lithium niobate and other lithium niobium oxides by rf magnetron sputtering *J. Appl. Phys.* **72** 4336–43

[25] Ievlev V, Sumets M, Kostyuchenko A and Bezryadin N 2013 Dielectric losses and ac conductivity of Si-LiNbO$_3$ heterostructures grown by the RF magnetron sputtering method *J. Mater. Sci. Mater. Electron.* **24** 1651–7

[26] Shandilya S, Tomar M, Sreenivas K and Gupta V 2009 Purely hopping conduction in c-axis oriented LiNbO$_3$ thin films *J. Appl. Phys.* **105** 94105

[27] Graça M P F, Prezas P R, Costa M M and Valente M A 2012 Structural and dielectric characterization of LiNbO$_3$ nano-size powders obtained by Pechini method *J. Sol-Gel Sci. Technol.* **64** 78–85

[28] Shandilya S, Tomar M, Sreenivas K and Gupta V 2009 Structural and interfacial defects in c-axis oriented LiNbO$_3$ thin films grown by pulsed laser deposition on Si using Al:ZnO conducting layer *J. Phys. D. Appl. Phys.* **42** 95303

[29] Elliott S R and Owens A P 1989 The diffusion-controlled relaxation model for ionic transport in glasses *Philos. Mag. Part B* **60** 777–92

[30] Elliott S R 1994 Frequency-dependent conductivity in ionically and electronically conducting amorphous solids *Solid State Ionics* **70–71** 27–40

[31] Casey H C, Cho A Y, Lang D V, Nicollian E H and Foy P W 1979 Investigation of heterojunctions for MIS devices with oxygen-doped $Al_x Ga_{1-x}$ As on n -type GaAs *J. Appl. Phys.* **50** 3484–91

[32] Nicollian E H and Goetzberger A 1967 The $Si-SiO_2$ Interface—electrical properties as determined by the metal-insulator-silicon conductance technique *Bell Syst. Tech. J.* **46** 1055–133

[33] Donnerberg H, Tomlinson S M, Catlow C R A and Schirmer O F 1989 Computer-simulation studies of intrinsic defects in $LiNbO_3$ crystals *Phys. Rev. B* **40** 11909–16

[34] Chen R H, Chen L-F and Chia C-T 2007 Impedance spectroscopic studies on congruent $LiNbO_3$ single crystal *J. Phys. Condens. Matter* **19** 86225

[35] Dhar A, Singh N, Singh R K and Singh R 2013 Low temperature dc electrical conduction in reduced lithium niobate single crystals *J. Phys. Chem. Solids* **74** 146–51

[36] Akhmadullin I S, Golenishchev-Kutuzov V A, Migachev S A and Mironov S P 1998 Low-temperature electrical conductivity of congruent lithium niobate crystals *Phys. Solid State* **40** 1190–2

[37] Dhar A and Mansingh A 1990 Polaronic hopping conduction in reduced lithium niobate single crystals *Philos. Mag. Part B* **61** 1033–42

[38] Schirmer O F, Imlau M, Merschjann C, Schoke B and D 2009 Electron small polarons and bipolarons in $LiNbO_3$ *J. Phys. Condens. Matter* **21** 123201

[39] Bergmann G 1968 The electrical conductivity of $LiNbO_3$ *Solid State Commun.* **6** 77–9

[40] Jorgensen P J and Bartlett R W 1969 High temperature transport processes in lithium niobate *J. Phys. Chem. Solids* **30** 2639–48

[41] Bollmann W 1977 The origin of photoelectrons and the concentration of point defects in $LiNbO_3$ Crystals *Phys. Status Solidi* **40** 83–91

[42] Ievlev V, Shur V, Sumets M and Kostyuchenko A 2013 Electrical properties and local domain structure of $LiNbO_3$ thin film grown by ion beam sputtering method *Acta Metall. Sin. (English Lett.)* **26** 630–4

[43] Kiselev D A, Zhukov R N, Bykov A S, Voronova M I, Shcherbachev K D, Malinkovich M D and Parkhomenko Y N 2014 Effect of annealing on the structure and phase composition of thin electro-optical lithium niobate films *Inorg. Mater* **50** 419–22

[44] Simões A Z, Zaghete M A, Stojanovic B D, Gonzalez A H, Riccardi C S, Cantoni M and Varela J A 2004 Influence of oxygen atmosphere on crystallization and properties of $LiNbO_3$ thin films *J. Eur. Ceram. Soc.* **24** 1607–13

[45] Fleetwood D M, Winokur P S, Reber R A, Meisenheimer T L, Schwank J R, Shaneyfelt M R and Riewe L C 1993 Effects of oxide traps, interface traps, and 'border traps' on metal-oxide-semiconductor devices *J. Appl. Phys.* **73** 5058–74

[46] Lai S K and Young D R 1981 Effects of avalanche injection of electrons into silicon dioxide —generation of fast and slow interface states *J. Appl. Phys.* **52** 6231–40

[47] Arnold D, Cartier E and DiMaria D J 1994 Theory of high-field electron transport and impact ionization in silicon dioxide *Phys. Rev. B* **49** 10278–97

[48] Cartier E 1998 Characterization of the hot-electron-induced degradation in thin SiO_2 gate oxides *Microelectron. Reliab.* **38** 201–11

[49] Warren W L and Lenahan P M 1987 Fundamental differences between thick and thin oxides subjected to high electric fields *J. Appl. Phys.* **62** 4305–8

[50] Warren W L and Lenahan P M 1986 Electron spin resonance study of high field stressing in metal-oxide-silicon device oxides *Appl. Phys. Lett.* **49** 1296–8

[51] Ievlev V, Sumets M and Kostyuchenko A 2013 Conduction mechanisms in Si-LiNbO$_3$ heterostructures grown by ion-beam sputtering method *J. Mater. Sci.* **48** 1562–70

[52] Ievlev V, Sumets M, Kostyuchenko A, Ovchinnikov O, Vakhtel V and Kannykin S 2013 Band diagram of the Si-LiNbO$_3$ heterostructures grown by radio-frequency magnetron sputtering *Thin Solid Films* **542** 289–94

[53] Buné A V and Pashkov V A 1986 Electron-drift mobility in lithium niobate crystals *Fiz. Tverd. Tela* **28** 3024–7

[54] Kireev P 1978 *Semiconductor Physics* (Moscow: Mir Publishers)

[55] Iyevlev V, Sumets M and Kostyuchenko A 2012 Current-voltage characteristics and impedance spectroscopy of LiNbO$_3$ films grown by RF magnetron sputtering *J. Mater. Sci. Mater. Electron.* **23** 913–20

[56] Ievlev V M, Sumets M P and Kostyuchenko A V 2012 Effect of thermal annealing on electrical properties of Si-LiNbO$_3$ *Mater. Sci. Forum* **700** 53–7

[57] Lampert M A and Mark P 1970 *Current Injection in Solids* (New York: Academic)

[58] Akazawa H and Shimada M 2007 Mechanism for LiNb$_3$O$_8$ phase formation during thermal annealing of crystalline and amorphous LiNbO$_3$ thin films *J. Mater. Res.* **22** 1726–36

[59] Elliott S R 1987 A.c. conduction in amorphous chalcogenide and pnictide semiconductors *Adv. Phys.* **36** 135–217

[60] Ching W Y, Gu Z-Q and Xu Y-N 1994 First-principles calculation of the electronic and optical properties of LiNbO$_3$ *Phys. Rev. B* **50** 1992–5

[61] Schmidt W G, Albrecht M, Wippermann S, Blankenburg S, Rauls E, Fuchs F, Rödl C, Furthmüller J and Hermann A 2008 LiNbO$_3$ ground- and excited-state properties from first-principles calculations *Phys. Rev. B* **77** 35106

[62] Yang J Y, Li W S, Li H, Sun Y, Dou R F, Xiong C M, He L and Nie J C 2009 Grain size dependence of electrical and optical properties in Nb-doped anatase TiO$_2$ *Appl. Phys. Lett.* **95** 213105

[63] Satapathy S, Mukherjee C, Shaktawat T, Gupta P K and Sathe V G 2012 Blue shift of optical band-gap in LiNbO$_3$ thin films deposited by sol–gel technique *Thin Solid Films* **520** 6510–4

[64] Dhar A and Mansingh A 1990 Optical properties of reduced lithium niobate single crystals *J. Appl. Phys.* **68** 5804–9

[65] Sumets M, Ovchinnikov O, Ievlev V and Kostyuchenko A 2017 Optical band gap shift in thin LiNbO$_3$ films grown by radio-frequency magnetron sputtering *Ceram. Int.* **43** 13565–8

[66] Thierfelder C, Sanna S, Schindlmayr A and Schmidt W G 2010 Do we know the band gap of lithium niobate? *Phys. Status Solidi* **7** 362–5

[67] Fox M 2010 *Optical Properties of Solids* (New York: Oxford University Press)

[68] Bhatt R, Bhaumik I, Ganesamoorthy S, Karnal A K, Swami M K, Patel H S and Gupta P K 2012 Urbach tail and bandgap analysis in near stoichiometric LiNbO$_3$ crystals *Phys. Status Solidi* **209** 176–80

[69] Wemple S H 1965 Some transport properties of oxygen-deficient single-crystal potassium tantalate (KTaO$_3$) *Phys. Rev.* **137** A1575–82

[70] Li X, Kong Y, Liu H, Sun L, Xu J, Chen S, Zhang L, Huang Z, Liu S and Zhang G 2007 *Origin of the Generally Defined Absorption Edge of Non-stoichiometric Lithium Niobate Crystals* vol 141

[71] Jiangou Z, Shipin Z, Dingquan X, Xiu W and Guanfeng X 1992 Optical absorption properties of doped lithium niobate crystals *J. Phys. Condens. Matter* **4** 2977–83

[72] Kitaeva G K, Kuznetsov K A, Penin A N and Shepelev A V 2002 Influence of small polarons on the optical properties of $Mg:LiNbO_3$ crystals *Phys. Rev. B* **65** 54304

[73] Sumets M, Dannangoda G C C, Kostyuchenko A, Ievlev V, Dybov V and Martirosyan K S S 2017 Temperature transition of p- to n-type conduction in the $LiNbO_3/Nb_2O_5$ polycrystalline films *Mater. Chem. Phys.* **191** 35–44

[74] Tan S, Gilbert T, Hung C-Y, Schlesinger T E and Migliuolo M 1996 Sputter deposited c-oriented $LiNbO_3$ thin films on SiO_2 *J. Appl. Phys.* **79** 3548

[75] Orton J W and Powell M J 1980 The Hall effect in polycrystalline and powdered semiconductors *Reports Prog. Phys.* **43** 1263–307

[76] Wilkinson A P, Cheetham A K and Jarman R H 1993 The defect structure of congruently melting lithium niobate *J. Appl. Phys.* **74** 3080

[77] Lim D, Jang B, Moon S, Won C and Yi J 2001 Characteristics of $LiNbO_3$ memory capacitors fabricated using a low thermal budget process *Solid. State. Electron.* **45** 1159–63

[78] Ling C H, Fisher J H and Anderson J C 1972 Carrier mobility and field effect in thin indium antimode films *Thin Solid Films* **14** 267–88

[79] Greener E H, Whitmore D H and Fine M E 1961 Electrical conductivity of near-stoichiometric α-Nb_2O_5 *J. Chem. Phys.* **34** 1017–23

[80] Popović R S 2004 *Hall Effect Devices* (Bristol and Philadelphia: Institute of Physics Publishing)

[81] Wang X, Liu X, Bo F, Chen S, Chen J, Kong Y, Xu J and Zhang G 2015 Photo-Hall effect in highly Mg-doped lithium niobate crystals *Appl. Phys. Lett.* **107** 191102

IOP Publishing

Lithium Niobate-Based Heterostructures
Synthesis, properties and electron phenomena
Maxim Sumets

Appendix A

Cell parameters and powder x-ray diffraction data of $LiNbO_3$ [1], $LiNb_3O_8$ [2], Li_3NbO_4 [3] and Nb_2O_5 [4]

$LiNbO_3$

Space group: R3c

Cell parameters: $a = 5.147$ Å, $b = 5.147$ Å, $c = 13.856$ Å, $\alpha = 90.0°$, $\beta = 90.0°$, $\gamma = 120.0°$.

2Θ (deg)	d-spacing (Å)	Intensity	h	k	l
23.715	3.7488	999	0	1	2
32.713	2.7353	349	1	0	4
34.83	2.5737	204	1	1	0
38.968	2.3094	34	0	0	6
40.073	2.2482	66	1	1	3
42.573	2.1218	105	2	0	2
48.529	1.8744	154	0	2	4
53.249	1.7188	209	1	1	6
54.844	1.6726	4	2	1	1
56.133	1.6372	120	1	2	2
56.995	1.6144	62	0	1	8
61.112	1.5152	107	2	1	4
62.448	1.4859	86	3	0	0
64.693	1.4397	2	1	2	5
68.557	1.3676	37	2	0	8
71.203	1.3232	32	1	0	10
73.536	1.2869	22	2	2	0
73.792	1.283	12	2	1	7

(*Continued*)

76.111	1.2496	28	3	0	6
76.834	1.2396	3	2	2	3
77.437	1.2315	2	1	3	1
78.524	1.2171	47	3	1	2
79.256	1.2077	40	1	2	8
81.775	1.1768	17	0	2	10
82.831	1.1644	30	1	3	4
83.686	1.1547	6	0	0	12
86.034	1.1291	1	3	1	5
86.508	1.1241	35	2	2	6
88.864	1.1003	16	0	4	2

LiNb$_3$O$_8$

Space group: P2$_1$/a

Cell parameters: $a = 15.2620$ Å, $b = 5.0330$ Å, $c = 7.4570$ Å, $\alpha = 90.0°$, $\beta = 107.3°$, $\gamma = 90.0°$.

2Θ (deg)	d-spacing (Å)	Intensity	h	k	l
12.15	7.2842	11.46	2	0	0
14.58	6.0760	22.18	−2	0	1
19.87	4.4686	13.95	2	0	1
21.46	4.1407	29.88	2	1	0
21.62	4.1095	39.21	0	1	1
22.94	3.8760	8.41	−2	1	1
23.9	3.7234	15.5	−4	0	1
24.34	3.6564	26.63	−2	0	2
24.44	3.6421	27.24	4	0	0
25.02	3.5591	11.53	0	0	2
26.68	3.3416	6.64	2	1	1
29.4	3.0380	6.91	−4	0	2
29.85	2.9933	2	−4	1	1
30.21	2.9581	100	−2	1	2
30.29	2.9506	99.26	4	1	0
30.73	2.9097	5.23	4	0	1
30.77	2.9059	1.06	0	1	2
31.08	2.8773	19.91	2	0	2
34.48	2.6009	2.39	−4	1	2
35.68	2.5165	23.69	0	2	0
35.95	2.4979	26.25	2	1	2
36.15	2.4846	3.13	−2	0	3
37.02	2.4281	3.19	6	0	0
38.14	2.3597	10.38	−6	0	2
38.63	2.3310	1.49	−4	0	3
38.73	2.3250	3.38	−2	2	1

39.74	2.2684	2.57	−6	1	1
40.49	2.2279	2.38	−2	1	3
40.63	2.2205	7.46	−3	2	1
40.81	2.2109	7.98	−1	1	3
40.94	2.2046	6.42	5	1	1
41.17	2.1927	2	2	2	1
42.1	2.1462	2.9	0	1	3
43.4	2.0850	4.44	−4	2	1
43.49	2.0808	1.27	2	0	3
43.66	2.0730	7.31	−2	2	2
43.72	2.0704	7.54	4	2	0
44.07	2.0547	2.88	0	2	2
44.36	2.0421	2.94	4	1	2
46.61	1.9487	4.25	6	1	1
46.88	1.9380	2.83	−4	2	2
47.27	1.9229	1.33	2	1	3
47.72	1.9059	1.15	−8	0	1
47.79	1.9034	2.48	4	2	1
48.03	1.8942	13.51	2	2	2
48.45	1.8789	4.58	−6	1	3
49.88	1.8282	6.12	−4	0	4
50.09	1.8210	5.9	8	0	0
51.26	1.7824	2.14	−8	1	1
51.34	1.7795	13.46	0	0	4
51.5	1.7746	13.57	6	0	2
51.7	1.7681	3.18	−2	2	3
51.88	1.7623	1	4	0	3
52.36	1.7473	3.33	6	2	0
52.4	1.7461	3.64	−8	1	2
52.41	1.7457	3.38	−2	1	4
53.21	1.7213	32.06	−6	2	2
53.31	1.7183	20.91	−4	1	4
53.51	1.7124	21.55	8	1	0
53.59	1.7101	1.91	−4	2	3
54.71	1.6777	1.17	0	1	4
54.86	1.6736	1.14	6	1	2
56.27	1.6349	1.43	2	3	0
56.34	1.6329	2.11	0	3	1
57.31	1.6076	1.66	−6	1	4
57.47	1.6036	1.5	2	2	3
60.74	1.5248	12.92	−2	3	2
60.79	1.5238	12.67	4	3	0
60.98	1.5193	1.27	−8	2	1

(*Continued*)

(Continued)

2Θ (deg)	d-spacing (Å)	Intensity	h	k	l
62.58	1.4844	1.17	−4	0	5
62.83	1.4791	4.76	−4	2	4
63.01	1.4753	4.71	8	2	0
64.03	1.4542	10.33	−8	1	4
64.09	1.4530	11.77	0	2	4
64.16	1.4515	10.24	−10	1	2
64.22	1.4503	12.11	6	2	2
64.27	1.4493	15.64	2	3	2
64.56	1.4435	1.08	4	2	3
64.8	1.4387	1.96	4	0	4
65.33	1.4283	1.1	6	1	3
66.46	1.4067	1.04	−6	2	4
66.95	1.3976	1.22	8	1	2
67.74	1.3833	11.12	4	1	4
68.49	1.3699	1.85	0	1	5
70.25	1.3399	1.02	−10	0	4
72.71	1.3005	1.78	−8	2	4
72.84	1.2985	1.97	−10	2	2
73.27	1.2920	1.47	−6	3	3
74.17	1.2785	1.22	−4	2	5
75.57	1.2582	2.71	0	4	0
75.75	1.2556	1.63	−10	2	3
76.42	1.2463	1.2	−8	3	2
77.17	1.2361	1.59	−4	3	4
77.33	1.2339	1.7	8	3	0
78.47	1.2188	2.56	−6	0	6
78.84	1.2140	2.73	12	0	0
80.32	1.1953	5.51	−2	1	6
80.59	1.1920	1.02	−4	4	1
80.65	1.1914	5.74	10	1	2
80.78	1.1898	1.1	−2	4	2
80.82	1.1893	1.09	4	4	0
81.36	1.1827	1.47	−10	2	4
81.6	1.1799	4.14	−12	0	4
83.94	1.1528	2.15	2	4	2
85.52	1.1355	1	−8	1	6
86.41	1.1260	3.17	−8	3	4
86.53	1.1248	3.2	−10	3	2
87.07	1.1192	1.47	2	0	6
87.27	1.1171	1.48	8	0	4
87.95	1.1103	5.86	−6	4	2
89.31	1.0969	4.4	−6	2	6
89.67	1.0934	4.66	12	2	0

89.76	1.0925	1.97	2	1	6
89.8	1.0921	4.18	4	3	4
89.96	1.0906	2.08	8	1	4

Li$_3$NbO$_4$

Space group: I23

Cell parameters: a = 8.429 Å, b = 8.429 Å, c = 8.429 Å, α = 90.0°, β = 90.0°, γ = 90.0°.

2Θ (deg)	d-spacing (Å)	Intensity	h	k	l
14.86	5.9602	94.81	1	1	0
21.08	4.2145	4.47	2	0	0
25.89	3.4411	100	2	1	1
33.62	2.6655	13.74	3	0	1
33.62	2.6655	18.05	3	1	0
36.94	2.4332	6.93	2	2	2
40.02	2.2527	6.67	3	2	1
40.02	2.2527	9.75	3	1	2
42.92	2.1072	45.27	4	0	0
45.66	1.9867	8.9	4	1	1
45.66	1.9867	10.67	3	3	0
50.81	1.7971	1.22	3	3	2
53.24	1.7206	3.58	4	2	2
55.6	1.6531	6.39	4	3	1
55.6	1.6531	4.97	4	1	3
55.6	1.6531	1.33	5	0	1
60.13	1.5389	13.25	5	2	1
60.13	1.5389	11.37	5	1	2
62.31	1.4901	20.2	4	4	0
64.46	1.4456	5.75	4	3	3
66.56	1.4048	1.07	6	0	0
68.64	1.3674	2.14	5	3	2
68.64	1.3674	4.39	6	1	1
68.64	1.3674	3.2	5	2	3
72.7	1.3006	1.37	5	4	1
72.7	1.3006	1.04	5	1	4
74.7	1.2707	3.13	6	2	2
76.68	1.2428	1.95	6	3	1
76.68	1.2428	1.54	6	1	3
78.64	1.2166	4.97	4	4	4
80.59	1.192	1.33	7	1	0

(*Continued*)

2Θ (deg)	d-spacing (Å)	Intensity	h	k	l
84.46	1.147	1.48	6	3	3
84.46	1.147	6.13	5	5	2
88.3	1.1068	1.82	7	3	0
88.3	1.1068	1.57	7	0	3

Nb₂O₅

Space group: C2/c (15)

Cell parameters: $a = 21.74$ Å, $b = 4.883$ Å, $c = 5.561$ Å, $\alpha = 90.0°$, $\beta = 105.02°$, $\gamma = 90.0°$.

2Θ (deg)	d-spacing (Å)	Intensity	h k l	Remarks
14.4	6.1523	214	2 0 0	
19.5	4.5386	2	1 1 0	
24.4	3.6425	999	−1 1 1	
26.9	3.314	715	1 1 1	
29.0	3.0761	566	4 0 0	
29.9	2.9811	927	−3 1 1	
32.7	2.7347	17	−2 0 2	
33.3	2.6854	378	0 0 2	
35.8	2.5033	331	3 1 1	
36.8	2.4415	178	0 2 0	
38.3	2.3466	76	−4 0 2	
39.7	2.2693	12	2 2 0	
39.9	2.2561	3	2 0 2	
40.6	2.22	99	−5 1 1	Multiply indexed line.
40.6	2.22	99	1 1 2	Multiply indexed line.
41.0	2.1976	59	5 1 0	
41.6	2.1659	39	−2 2 1	
44.1	2.0507	55	6 0 0	
44.8	2.0222	1	2 2 1	
46.9	1.9342	4	−5 1 2	
47.5	1.9123	221	4 2 0	
48.2	1.8866	32	5 1 1	
48.3	1.882	53	−6 0 2	
48.7	1.8663	4	3 1 2	
50.0	1.8212	11	−2 2 2	
50.5	1.8065	302	0 2 2	Doubly indexed line.
50.5	1.8065	302	4 0 2	Doubly indexed line.
53.1	1.7225	118	−1 1 3	
53.3	1.7163	67	4 2 1	
53.6	1.7079	157	−3 1 3	
53.8	1.7016	271	−7 1 1	
54.1	1.6918	242	−4 2 2	
55.4	1.657	6	2 2 2	

55.5	1.6539	4	7 1 0	
57.0	1.6136	71	1 3 0	Doubly indexed line.
57.0	1.6136	71	1 1 3	Doubly indexed line.
57.7	1.5948	3	−6 2 1	
57.8	1.5922	3	−7 1 2	
58.4	1.5783	8	−5 1 3	
58.7	1.5703	45	6 2 0	
59.2	1.56	47	−1 3 1	
60.1	1.538	42	8 0 0	
60.2	1.5354	26	5 1 2	
60.4	1.5311	52	1 3 1	
61.1	1.5146	36	−8 0 2	
62.1	1.4939	84	−3 3 1	
62.2	1.4905	82	−6 2 2	
62.6	1.4822	112	7 1 1	
63.8	1.4578	38	6 0 2	
64.1	1.4512	91	4 2 2	
64.5	1.4437	8	0 2 3	
64.8	1.4374	75	3 1 3	
65.0	1.4325	40	6 2 1	
65.6	1.4212	127	3 3 1	
66.5	1.4043	3	−1 3 2	
66.9	1.3963	82	−7 1 3	
68.1	1.3755	1	−3 3 2	
68.5	1.3673	42	−4 0 4	
68.8	1.3633	28	−5 3 1	Doubly indexed line.
68.8	1.3633	28	1 3 2	Doubly indexed line.
69.0	1.3595	6	−9 1 1	
70.0	1.3427	38	0 0 4	
70.5	1.334	4	−3 1 4	Doubly indexed line.
70.5	1.334	4	−8 2 1	Doubly indexed line.
71.4	1.3194	5	−9 1 2	
71.5	1.3172	7	−6 2 3	Doubly indexed line.
71.5	1.3172	7	9 1 0	Doubly indexed line.
72.6	1.3013	30	8 2 0	
73.5	1.287	40	−8 2 2	Multiply indexed line.
73.5	1.287	40	−6 0 4	Multiply indexed line.
74.2	1.2761	1	7 1 2	
74.4	1.2736	4	5 3 1	
74.8	1.2673	1	3 3 2	
75.8	1.2532	5	5 1 3	
75.9	1.2516	8	6 2 2	
76.2	1.2477	23	−10 0 2	

(*Continued*)

(Continued)

2Θ (deg)	d-spacing (Å)	Intensity	h k l	Remarks
77.5	1.2304	8	10 0 0	
78.2	1.2207	22	0 4 0	
78.7	1.2141	39	−3 3 3	
78.9	1.2119	35	−7 3 1	
79.3	1.2066	14	−2 2 4	Doubly indexed line.
79.3	1.2066	14	8 0 2	Doubly indexed line.
80.0	1.1974	3	2 4 0	
80.4	1.193	32	−4 2 4	
80.6	1.1903	19	0 4 1	
81.4	1.1814	6	−2 4 1	
81.6	1.1788	23	1 3 3	
81.8	1.1765	49	0 2 4	
82.0	1.1733	40	−8 0 4	
82.3	1.1704	13	−7 3 2	
82.8	1.1649	6	−5 3 3	
83.7	1.1545	1	3 1 4	
84.3	1.1473	1	5 3 2	
85.2	1.1381	13	−6 2 4	
85.5	1.1346	40	4 4 0	
85.7	1.1322	20	−4 4 1	
86.1	1.128	20	4 0 4	
86.4	1.1246	52	−11 1 1	Doubly indexed line.
86.4	1.1246	52	7 3 1	Doubly indexed line.
87.4	1.1147	3	−2 4 2	
87.7	1.1113	49	0 4 2	Doubly indexed line.
87.7	1.1113	49	−10 2 2	Doubly indexed line.
88.4	1.1046	23	3 3 3	
89.0	1.0988	32	10 2 0	
89.9	1.0896	33	7 1 3	Doubly indexed line.
89.9	1.0896	33	11 1 0	Doubly indexed line.

References

[1] Abrahams S C and Marsh P 1986 Defect structure dependence on composition in lithium niobate *Acta Crystallogr. Sect. B Struct. Sci.* **42** 61–8

[2] Lundberg M 1971 The crystal structure of $LiNb_3O_8$ *Acta Chem. Scand.* **25** 3337–46

[3] Grenier J, Martin C and Durif A 1964 Etude cristallographique des orthoniobates et orthotantalates de lithium *Bull. la Soc. Fr. Mineral. Cristallogr.* **87** 316–20

[4] Ercit T S 1991 Refinement of the structure of Nb_2O_5 and its relationship to the rutile and thoreaulite structures *Mineral. Petrol.* **43** 217–23

CPSIA information can be obtained
at www.ICGtesting.com
Printed in the USA
BVHW011152220721
612410BV00022B/364

9 780750 317702